AF183043

Manual Eight

Display Devices

The McGraw-Hill OPERATION UPDATE Series in
INTEGRATED CIRCUIT TECHNOLOGY AND APPLICATIONS

Manual Eight

Display Devices

New York
St. Louis
San Francisco
Auckland
Düsseldorf
Johannesburg
Kuala Lumpur
London
Mexico
Montreal
New Delhi
Panama
Paris

McGRAW-HILL BOOK COMPANY

São Paulo
Singapore
Sydney
Tokyo
Toronto

Library of Congress Cataloging in Publication Data

Main entry under title:
Integrated circuit technology and applications.

(The McGraw-Hill operation update series)
Includes an introductory vol. with special title: Master outline and guide for The McGraw-Hill operation update series in integrated circuit technology and applications and a concluding vol. with special title: Problems and answers.
Originally published under title: Integrated-circuit technology, analog and logic circuit design, memory and display devices.
Includes bibliographies and indexes.
Partial Contents: manual 1. Grebene, A.B. Integrated circuit processes.—manual 2. Grebene, A. B. Integrated circuit devices and components.—manual 3. Millman, J. and Halkias, C.C. Operational amplifiers. [etc.]

1. Integrated circuits. 2. Microelectronics.
I. McGraw-Hill Book Company.
TK7874.1543 1978 621.381′73′08 78-3237
ISBN 0-07-079212-7 (Set)
ISBN 0-07-044668-7 (Manual 8)

New reprint 1979.

1234567890 EBEB 7865432109

1221 Avenue of the Americas
New York, New York 10020

Display Devices

Reading Assignment

Selected invited papers from the July 1973 *Proceedings of the IEEE,* Special Issue on New Materials for Display Devices:

E. I. Gordon and L. K. Anderson, New Display Technologies—An Editorial Viewpoint.
M. G. Craford and W. O. Groves, Vapor Phase Epitaxial Materials for LED Applications.
L. T. Creagh, Nematic Liquid Crystal Materials for Displays.

Purpose

In the integrated analog and digital electronics considered in earlier manuals, the principal material employed is always silicon, so that processing procedures (such as epitaxial growth, oxidation, and diffusion) are similar for different types of integrated circuits. In displays, a variety of different materials with different technologies are available, each with particular advantages and disadvantages. In this manual, an overview of display technologies is presented and two of the newer technologies, light emitting diodes and liquid crystals, are described in some depth. The main purpose is to give an overall perspective in the display area, while focusing on two competing technologies of increasing use in electronic equipments.

Many applications of integrated electronics involve some types of information displays. As the size, cost, and reliability of the integrated electronics described in previous manuals improves, displays become more important in the overall equipment design. Thus an understanding of the advantages and limitations of these expanding technologies becomes of increasing importance to the equipment designer.

Summary

In the first paper by Gordon and Anderson, a variety of displays, such as cathode-ray tubes (CRT), plasma (both dc and ac excited), electroluminescence, light-emitting diodes (LED), and liquid crystals are summarized with emphasis on areas of application, advantages, and disadvantages. The difficult task of classifying displays is attempted from a variety of perspectives, with emphasis on the difference between *(a)* light-emitting and light-controlling displays and *(b)* beam-addressed and matrix-addressed displays. The variety of display technologies are described in this classification scheme. For example, light-emitting diodes are light-emitting and matrix-addressed displays, while liquid crystals are light-controlling, matrix-addressed displays. Both of these displays have drive requirements compatible with silicon integrated circuits, a key advantage for acceptance in modern electronic equipment.

In the second paper by Craford and Groves, gallium arsenide phosphide (GaAsP) and gallium phosphide (GaP) light-emitting diode technology are described, with particular emphasis on the commercially-dominant, vapor phase epitaxial (VPE) growth (as opposed to liquid phase epitaxy). The paper provides a good summary of VPE growth, device fabricating material, and device characterization, detailed comparison of VPE and LPE devices, and economic projections.

In the third paper by Creagh, a similar but less detailed treatment of nematic liquid crystals is presented. The organic liquid crystal technology is vastly different from the semiconductor technology used in silicon integrated circuits and light-emitting diodes. This article does not address material aspects in detail but focuses on display device aspects. Particular attention should be focused on the limited temperature range of liquid crystals (both operating and storage) and a comparison of the various modes of operation (AC drive dynamic scattering, DC drive dynamic scattering, and various field effects).

Additional Reading

The individual papers used in the reading assignment are well referenced. In addition, the *Proceedings of the IEEE,*

Vol. 61, No. 7, July 1973, Special Issue on New Materials for Display Devices, from which the reading assignment was selected is highly recommended for more depth. References listed below are to other articles in this Special Issue:

E. E. Loebner, The Future of Electroluminescent Solids in Displays Applications. An extensive, thorough review of electroluminescent solids including those in the research stage at present. Contains a detailed discussion of human color vision.

E. Schlam, Electroluminescent Phosphors. Focuses on properties of the zinc sulfide family. Good treatment of material and devices capabilities.

S. Larach and A. E. Hardy, Cathode-Ray-Tube Phosphors: Principles and Applications. Review of CRT development and present status of most widely employed display device.

Z. Van Gelder and M. M. M. P. Mattheij, Principles and Techniques in Multicolor DC Gas Discharge Displays. Summay of techniques of extending DC gas discharges into multicolor data and video displays.

New Display Technologies—An Editorial Viewpoint

EUGENE I. GORDON AND LAWRENCE K. ANDERSON

Abstract—**Present-day display technologies, both exploratory and in common use, are reviewed in outline form with emphasis on areas of applicability, advantages, and disadvantages. The comments reflect the authors' own bias, but the basic ideas probably reflect the thinking of most of the members of the display community.**

I. Introduction

THE PURSUIT of new display technology has been characterized by many starters but very, very few finishers. Part of the reason is the hazard common to all developments that attempt to satisfy a market already well served by an existing capability: the current technology *always* has the inside track, for reasons which, while not always rational, are usually compelling. The most significant factor is the long development cycle from invention to marketed product. The rate of change in the technological, economic, and customer environment is so great that the original motivation is frequently lost. Newer, more promising approaches present themselves. In the meantime, existing technology evolves and adapts to the changing environment.

In the case of display, the range of possibilities for producing and controlling light is enormous and many attractive opportunities present themselves. This has the effect of spreading resources among many subthreshold efforts. Prospects for new display technologies are further clouded by the fact that for many of the most important applications there exists a device, the familiar CRT, that has long provided a versatile, elegant, functional, economical, and largely satisfactory solution.

Recently, however, there have been increasing pressures to develop new display technologies to satisfy a wide range of requirements, including, in many cases, the requirement that the display have all the versatility of a CRT but be lower cost, more compact, more reliable, and consume less power. Although such a display is nonexistent, the pressures for it are very real. Furthermore, modern developments in large-scale silicon integrated circuits (SIC) have dramatically decreased the cost, bulk, and power consumption of all digital and most analog electronics. As this trend continues, most of the cost and bulk of terminal electronic apparatus is likely to be in the input and output transducers, e.g., displays, unless these are integrated as well. Even in image pickup devices, enormous technological strides are being made as a result of advances in silicon technology [1].

Another important impetus for change results from the increasing use of displays as an interface between humans and the computer. Here the emphasis is on digital address, linearity, intrinsic memory, and low bulk. Some applications particularly stress flat-panel and bright ambient viewing characteristics. Other applications require the display of only a small number of numeric or alphanumeric characters. Finally, the normal tendency to newness provides impetus to changing technology.

Manuscript received March 13, 1973.
The authors are with Bell Laboratories, Murray Hill, N. J. 07974.

Unfortunately, the microminiaturization techniques that have so dramatically lowered the cost of most electronic functions are fundamentally inapplicable to displays since a display must interface with a human, on a human-size scale. Thus in order to reduce display cost and bulk, much effort has gone into developing new, easily integrated, planar technologies amenable to batch fabrication, in the manner of silicon large-scale integration (LSI), but on a coarser scale. Included in such technologies are such developments as liquid crystals, plasma panels, light-emitting diodes (LED), thin-film electroluminescence, electrophoretic and various projection light-valve technologies. In this paper, we will summarize these major new display developments in outline form, and try to assess their applicability and relative advantages and disadvantages. This is a highly subjective procedure—hence the word "editorial" in our title—and perhaps the only justification for attempting it at all is to provide an informed, but possibly biased, point of view on a number of new technologies.

II. Classification of Displays

Any critical examination of display technologies must be done within a well-designed framework. From the user's point of view, the important parameters are 1) resolution (the number of picture elements displayed), 2) screen size (as measured by the number of people who are to view the display together), and 3) physical bulk.

In terms of resolution, displays can range all the way from simple one-element indicators to very high resolution displays with greater than a million picture elements. One arbitrary but useful scheme for classifying displays according to resolution is indicated in Table I.

Displays may also be classified according to the number of people who are expected to view the display together. Here it seems to make sense to distinguish only three classes—individual viewing, small group viewing, and large group viewing, as summarized in Table II.

Another important criterion from the users' point of view is the bulk of the display. In the case of a conventional console, such as a TV set, the bulk, i.e., the volume behind the screen, is well defined, and as this approaches zero we achieve our Utopian flat panel. Bulk is large, but less well defined in projection systems, where, in principle, the entire volume between projector and screen must be charged up against the display, but since one may have no good use for this space anyway, the question often becomes academic except in the case of rear projection screens. One can also have a low-bulk virtual-image display that can achieve large apparent size while maintaining brightness, when magnifying optics are incorporated between the display and the eye. The main penalty one pays for this sleight of hand, of course, is a correspondingly limited viewing angle so that the display is useful only for individual or very small group viewing.

One can also categorize displays, from the technologist's point of view, according to whether they are light *emitters* or

TABLE I

RESOLUTION

Display Categories	Number of Elements (N)
Indicators	$N=1$
Small displays	$1<N\lesssim 10^2$
Medium-sized displays	$10^2\lesssim N\lesssim 10^4$
Large displays	$10^4\lesssim N\lesssim 10^6$
High resolution	$10^6\lesssim N$

TABLE II

SIZE

Individual viewing	Wristwatch Calculator Meter Remote terminal Computer-aided design Picturephone® display
Small group viewing	Entertainment TV Video classroom Information bulletin
Large group viewing	Command displays Video conferencing Theater TV

light *controllers*. In the former case, the display is self-luminous, e.g., a CRT, LED, plasma panel, electroluminescent film; whereas, in the latter case, the viewing light is provided by an external source, such as a projection lamp, or just by the general ambient, and the display merely modulates this light, in transmission or reflection. Examples of displays in this category, which are often called *light valves*, are various film systems, liquid-crystal displays, cathodochromic tubes, various electrooptic light-valve systems, and the "eidophor" projection theater TV system.

The difference between light-emitting and light-controlling displays shows up, of course, in a myriad of engineering details, but it shows most fundamentally in power requirements. In a light-emitting display, by definition, power to generate the light by which it is viewed must be charged against the display. Depending on such circumstances as the required size and brightness, which in turn depend on the ambient light level with which the display must compete, this power can dominate the total power requirements. With light-valve displays, on the other hand, it is sometimes possible to force the ambient to provide all the viewing light, so that in principle the power requirements for these displays can approach zero, with obvious advantages in portable apparatus.

In terms of engineering detail, an important distinction can be made between beam-addressed displays and matrix-addressed displays. This has slightly different ramifications for light-emitting and light-controlling displays. In beam-addressed displays, as commonly understood, each element of the display must be addressed sequentially. It is supposed that the element is excited only during the address time. This is known as spot sequential operation. It imposes very definite, and often very severe, requirements on the individual display elements. In a beam-addressed light-emitting display having N picture elements for example, each element must be capable of responding, without saturation, to an input excitation N times greater than that required to achieve the same average brightness with steady-state excitation. While the period of optical emission may extend beyond the address time, this requirement often means that the individual display element must be capable of very high peak brightness. The phosphor in a CRT is in fact capable of the required peak brightness. The situation is different in matrix-addressed light-emitting displays. While spot sequential operation of these displays is possible, one has the option of using storage, typically in conjunction with line-at-a-time addressing, to reduce the peak brightness requirements. For line-at-a-time addressing, since the data to be displayed are almost always arriving in serial fashion, one must then, of course, provide series-to-parallel conversion with line storage—multilevel if gray scale is involved. In only a very few cases is full-frame storage required to achieve usable brightness.

In light-valve displays, the difference between beam and matrix addressing shows up in terms of a tradeoff between achievable contrast ratio and storage requirements. If the picture elements in an N-element display are addressed sequentially, and no storage at all is provided, and a time-averaged contrast ratio of C is required in the display, then the individual display elements must exhibit a peak contrast ratio of CN. This requirement is in addition to the necessity of providing adequate illumination to achieve a peak brightness N times the time-averaged brightness desired. These requirements are impossible to meet in practice, since even the best light-valve technologies show a peak contrast ratio of less than 50:1, whereas effective displays must generally have a time-averaged contrast ratio of greater than 10:1. Thus practical light-valve displays must invariably be designed to have full-frame storage. This requirement makes difficult the use of matrix-address techniques for all but low-resolution light-valve displays.

One of the major engineering distinctions that must be made between beam-addressed and matrix-addressed displays is in terms of the threshold requirement placed on the drive versus response characteristic of the individual display elements. In a beam-addressed display of N elements it is not particularly difficult to design the system so that the average drive on an element that is supposed to be off is less than that on an element that is supposed to be on by a factor of at least CN, where C is the desired contrast ratio, even when N is as large as 10^6. Thus the individual display element can have a linear drive-response characteristic, with no trace of a threshold, and still exhibit good contrast. In matrix-addressed schemes the necessary isolation of the "off" elements is generally impossible to achieve, so that the individual display element must show a highly nonlinear drive response characteristic, preferably with a well-defined threshold. The degree of nonlinearity required depends, of course, on the number of picture elements and the actual matrix-addressing scheme used. If the required nonlinearity must be provided by external circuit elements, such as semiconductor diodes, associated one-for-one with each picture element, the economics of the situation rapidly get out of hand for all but the smallest displays. Thus a major thrust of display research has been the search for simple, inexpensive display elements with intrinsic threshold.

Light-valve displays have received a great deal of attention in the recent past mainly because they offer solutions for bright ambient and projection viewing in addition to low power as has been mentioned. While matrix-address tech-

TABLE III

CRT DISPLAYS

Applications	Large displays Small and large group viewing Console, projection, and virtual image
Advantages	Bright Efficient Uniform All colors Gray scale Functional Low cost Reliability can be high (>20 000 h) Storage or memory is possible at expense of gray-scale capability
Disadvantages	Bulk[a] High voltage Memory is needed for gray scale or interactive graphics[b] Nondigital address

[a] Flat-panel CRT's are being developed but not all problems are solved.

[b] Frame memory has become relatively inexpensive and of high quality. Selective erase is possible.

TABLE IV

PLASMA (DC-EXCITED) DISPLAYS

Applications	Indicators, small, medium, and large displays Small group viewing Flat panel
Advantages	Bright (if display resolution is limited) Range of colors Gray scale Functional scanning Low cost per element Potential for long life
Disadvantages	High drive power Line memory required Limited number of scanning lines ($\lesssim 250$)

niques have proved useful for small and medium-sized light-valve displays, beam-address techniques are invariably used for the large and high-resolution displays.

The earliest light-valve displays, as exemplified by the eidophor system, were electron-beam addressed but most attention has been given to light addressing. Here the resolution elements are addressed by controlled points of light originating from a primary display. Hence the light-addressed light valve is appropriately characterized as an appliqué on a primary display [2]. The appliqué may provide storage, enhance the brightness or contrast of the displayed image, change color, or even allow color images from a monochrome display (even an IR primary display). The appliqué broadens the choice of primary displays that may be utilized for a given application. As an example, one might appliqué a liquid crystal–photoconductor light valve onto a plasma panel to achieve a digitally addressed flat-panel display useful for very bright ambient or projection viewing.

Comparison of light-addressed light-valve displays should not be made without also considering the associated primary display since the overall system performance, cost, reliability, etc., are what is relevant. While obvious, this point is frequently ignored.

These remarks have been very general, designed only to provide a framework within which we can examine somewhat more explicitly the features and shortcomings of today's crop of display technologies. This we will do in the next two sections, first for light-emitting displays, then for light-valve displays.

III. LIGHT-EMITTING DISPLAYS

A. Beam-Addressed

1) Electron-Beam-Addressed; Cathode-Ray Tube: The applications, advantages, and disadvantages of the CRT [3] are summarized in Table III. Most of the comments are believed to be noncontroversial. The reliability of CRT's used for the Bell System Picturephone® application has been well established by comprehensive life testing and results from careful choice of materials and processing. Increasing use of SIC's for the peripheral drive systems ensures increasing stability of the overall system. CRT's with inherent memory (storage tubes) usually sacrifice gray scale. However, excellent analog frame memory is becoming increasingly available at relatively low cost. The triode-type storage tube is an example of analog frame memory being used in many applications [4].

2) Laser-Beam-Addressed: The concept of scanning a visible laser across a screen to provide a directly viewed image is a superficially attractive one, especially since the required optical modulation and deflection technologies are themselves rather well developed [5]. And, indeed, very attractive full-color laser video displays large enough to fill a theater screen have been demonstrated [6]. At the present time, however, such displays are wholly impractical for the simple reason that the required visible laser power ($\sim$1 W/m^2 of screen area) cannot be achieved cheaply and efficiently enough. The scanned laser concept, at the moment, looks much more promising when used as a primary display to provide beam-addressing for a light-valve system, as discussed in Section IV.

B. Matrix-Addressed

1) The Digisplay (Northrop Corporation) is a CRT that provides a flat-panel geometry and digital address [7]. It has all of the other attributes of the CRT except functional scanning. Indeed, scanning from one point to the next implies switching upwards of 100 V across several hundred picofarads. At TV rates this requires high drive power per unit display area and poses severe requirements on the drive circuitry. Separation of the switching and modulation function allows good half-select and threshold characteristics and reasonable gray scale.

2) The self-scanning plasma panel (Burroughs Corporation) is described in Table IV [8]. While normally used for alphanumeric display, recent experiments indicate a high potential for use in TV display [9]–[11]. Colors are available from phosphors using photoexcitation from UV photons generated in the discharge [12] or direct excitation from low-energy discharge electrons [11]. Unfortunately, the achievable peak brightness is not adequate and some form of memory is required. Functional scanning limits this to line memory at most. The duty cycle is therefore equal to the reciprocal of the number of scanning lines. The required average brightness combined with the achievable peak brightness implies a minimum duty cycle of 0.004 and therefore a maximum of 250 scanning lines. As a primary display to be coupled to a light valve, its usefulness could be extended beyond present applications.

TABLE V

PLASMA (AC-EXCITED) DISPLAYS

Applications	Indicators, small, medium, and large displays Small and large group viewing Flat panel
Advantages	Bright Range of colors Inherent memory Threshold for switching Low cost per element Potential for long life Selective erase
Disadvantages	No gray scale Questionable compatibility with SIC Relatively high drive power Expensive overall display system

TABLE VI

ELECTROLUMINESCENT DISPLAYS

Applications	Small and medium-sized displays Individual, small, and large group viewing Flat panel
Advantages	Cost per element can be low Gray scale Various colors
Disadvantages	No internal memory No threshold Not compatible with SIC Uniformity can be problem Poor reliability

3) The ac plasma panel (Owens-Illinois) is described in Table V [13]. This panel is used strictly for binary level display. While techniques for displaying several discrete brightness levels have been demonstrated, it is not capable of true gray-scale display [14]. It has good threshold characteristics for matrix address and inherent long-term memory. The latter characteristic allows reasonable brightness. The "red" appearance of the display contains a considerable range of wavelengths and is more pleasant than typical LED reds. One of the characteristics shared with the dc panel is the relatively high uniformity of the display when operated in the binary mode.

At such time as the temporal and spatial variations in breakdown and sustain voltages are brought within limits so that modest requirements can be imposed on SIC drive circuitry, the cost picture and general utility of these displays will improve remarkably.

4) Electroluminescent displays are described in Table VI [15]. The concept of a thin-film flat panel using ac excitation has always been attractive. Unfortunately, the technology has proved to be exceedingly difficult and unyielding. Hope at the present time focuses on thin-film transistor logic, memory, and drive circuitry to bring the potential costs into line for digital display. Uniformity and reliability continue as major problem areas. Gray-scale display is even more challenging. At best this is a risky area for display development.

5) LED's are described in Table VII [16]. LED's have proved to be relatively important for indicators and small numeric and alphanumeric displays. They are in very tight competition with dc excited gas discharge displays and liquid crystal displays for the latter application particularly where cost and power are dominant criteria, e.g., hand-held calculators. LED's have a very good appearance and the closest compatibility with SIC drive circuitry but are limited in size. Liquid crystals have the lowest input power requirements but the poorest appearance and are not yet highly reliable. Gas-discharge displays have a good appearance, can be made quite large, and are reasonably reliable. They are least compatible with SIC drive circuits. The net result is that the three technologies will be in constant conflict, with LED's edged out of the competition for large or multicharacter ($\gtrsim 10$) displays. The precise crossover size will vary with time and will be highly dependent on the basic cost of the III–V substrates and the subtleties of the specific system design.

TABLE VII

LED DISPLAYS

Applications	Indicators and small displays Individual viewing Flat-panel and virtual image
Advantages	Bright Relatively efficient Red, yellow, and green SIC compatibility Small size Reliability ($>10^6$ h) Latching capability[a]
Disadvantages	Cost per element is fundamentally high and proportional to area

[a] Hardly necessary for small displays and almost never used.

In Table VII the comments concerning efficiency of the LED's will be challenged. Our evaluation is based on the efficiency of the associated drive circuitry and the pace of progress in the technology. Power efficiencies as high as 15 percent have been reported for GaP red material [17] although 2–5 percent is more typical. Green efficiency above one-half percent at low current density is in the offing.

IV. LIGHT-VALVE DISPLAYS

A. Beam-Addressed Light Valves

1) Electron-Beam-Addressed:

a) A typical example is the electrooptic birefringent light valve with KDP, e.g., the Philips "Titus" tube [18]. These devices are in an exploratory phase and it is difficult to evaluate their potential. Presumably they are intended to compete with projection CRT's.

b) The cathodochromic dark-trace tube [19] is not yet commercially important although it has been used in operational radar systems. The major drawback here is the relatively slow speed with which information of useful contrast may be written, although high-resolution gray-scale images may be stored for long periods. Improved tubes using sodalite targets appear to be well suited to slow-scan display applications.

c) One of the best known devices in this category is the oil-film diffractive light valve exemplified by the "eidophor" [20]. It is used solely for projection display. While it is bulky and relatively unreliable, it performs admirably in a real-time TV mode.

d) A related storage-display device for projection display of single TV frames is the Lumatron [21]. Unless there is a need for long-term readout and/or projection, the CRT combined with a triode storage tube is a more realistic and practical way to achieve the same function.

TABLE VIII

ERASABLE LIGHT-ADDRESSED LIGHT VALVES

Applications	Large and high-resolution displays Small and large group viewing Projection and virtual image
Advantages	High resolution Uniform Overall low cost per element Inherent memory Functional scanning
Disadvantages	Slow Useful only for relatively fixed displays Bulk

2) *Light-Beam-Addressed:* As described in the foregoing, these may be considered to be appliqués on a primary display. Examples are the following.

a) Laser-addressed film systems such as bismuth film, Dylux®, Kalvar, and dry silver which allow very high resolution, high-quality projection displays [2]. In the case of the thin metallic films (bismuth), no development is necessary and the image can be viewed as it is written. The same is true of the Dylux system. The Kalvar and dry silver systems require short-duration thermal development. Except for the Dylux system which loses detail by dye diffusion, the other systems are permanent and the film may be used as hard copy. The metal-film systems have a special advantage in that miniature, solid-state IR lasers may be used as the writing source. The others require visible or UV gas lasers. As lasers and the associated optics and deflectors improve, systems based on nonreusable film will find increasing use in facsimile and data retrieval displays. They should also prove useful in conferencing and remote classroom applications.

b) Erasable light-beam-addressed light valves form another important class of display elements, whose general properties are summarized in Table VIII. Many of these involve the use of a short-wavelength-sensitive photoconductive layer applied to an electrooptic medium. One such family of devices is the "ferpic" [22], [23] and "cerampic" [24] of Bell Laboratories and Sandia, respectively, in which a thin photoconductive film of CdS, CdS–ZnS, or an organic material, PVK, is applied to a thin plate of PLZT transparent ferroelectric ceramic. Another is the photoconductively accessed liquid-crystal light valve, such as that developed by Hughes [25], in which a thin-film photoconductor (CdS or ZnS) is combined with a cholesteric liquid-crystal cell to provide the storage light-valve function. Depending on the choice of liquid crystal, displays suitable for real-time TV or long-term storage can be fabricated. Another entry [26], based on the Pockels effect in KDP, has been dubbed "the Phototitus" because it is similar in principle to the Titus tube. It combines KDP with a photoconductor to produce light modulation by reflection. The Phototitus requires cooling and, because of its complexity, is more suited to coherent image processing or storage. Its performance is quite impressive, however. A related device is the Pockels effect light valve developed by Itek [27], in which the photoconductive and electrooptic functions are combined in a single, high-resistivity, monocrystalline material, such as ZnS or $Bi_{12}SiO_{20}$, which exhibits both a linear electrooptic effect and photoconductivity.

In all these displays, the local voltage drop across the active optical element, and hence its optical state, depends on the local light flux incident on the photoconductor. In many cases, the required light flux is sufficiently small (10^{-5}–10^{-4} J/cm^2 for the Hughes display) that nonlaser primary displays (e.g., a CRT) may be used to write in the image. When there exists a threshold for the writing beam intensity below which no write-in occurs, as in the case of the PLZT light valves, it is possible, with a relatively simple dichroic optical system, to arrange simultaneous write and project, just as in the case of the film light-valve systems discussed in subsection a) immediately preceding. Since the liquid-crystal system can operate at standard TV frame rates, the technology is potentially applicable to large-screen theater TV. Response at the high scan rates required for TV-like applications (~0.5 μs/picture element) is possible even though the response time of the liquid crystal is long (milliseconds) because of the combined effects of nonzero carrier lifetime in the photoconductor and capacitive storage across the liquid crystal cell. Much development work remains, however, to optimize various material parameters to obtain satisfactory contrast and gray scale.

Thermally written liquid-crystal light valves [28] offer a useful alternative to the photoconductively accessed devices discussed earlier. These light valves consist of a thin film of a cholesteric or smectic liquid crystal sandwiched between IR absorbing but visible transmitting indium–tin oxide electrodes. Write-in is accomplished with a focused low-power (few milliwatts) steered IR laser beam that is absorbed by the electrodes, locally heating the liquid crystal into the isotropic state. Heated regions of the liquid crystal cool down into a disordered highly scattering state and show up dark on a screen when the light valve is operated in an apertured (schlieren) optical system. Erasure is accomplished with an ac electric field, which reorients the individual liquid-crystal molecules. Local erasure is possible in smectic liquid crystals for which reorientation occurs only under simultaneous application of heat and electric field [29].

Thermally addressed liquid-crystal light valves offer simultaneous write and view with good contrast (>10:1), high resolution (>50 line pairs/mm), and low insertion loss (<1.5 dB) in an inherently simple storage device. While operation at TV frame rates does not seem practical, the device should find widespread use in a variety of graphics applications.

B. Matrix-Addressed Light Valves

As discussed in Section II, matrix-addressing of light valves imposes severe requirements on the properties of the individual display elements. As a bare minimum, drive voltages and currents must be low enough to allow easy interfacing with SIC's. Because of their modest drive requirements, in the past by far the largest activity has gone into liquid-crystal devices (for a recent review of liquid-crystal devices, see [30]), although electrochromic devices using thin films of transition metal compounds [31], colloidal dipole suspensions [32], and, more recently, electrophoretic suspensions [33] have also been demonstrated. Some general properties of the liquid-crystal displays are summarized in Table IX. Aside from appearance, properties of the dipole and electrophoretic suspensions are generally similar to those of the liquid-crystal devices except that lifetime is unproved and of even greater concern.

Until very recently, most liquid-crystal displays have

TABLE IX

LIQUID-CRYSTAL DISPLAYS

Matrix-Addressed

Applications	Small and medium-sized displays Individual viewing Flat panel
Advantages	Good contrast in bright ambient Low power SIC compatible Low cost per element Uniform
Disadvantages	Limited temperature range (−10°C to +60°C at best) Limited reliability (≤50 000 h ac, worse for dc) Requires memory (one memory element per pixel) Low switching speed (tradeoff in drive power and life versus speed)

used "dynamic scattering" [34], in which scattering is induced by ionic flow in a thin film of a moderate-resistivity nematic liquid crystal. This requirement for current flow had led to some concern over lifetime, although evidence is mounting that carefully prepared hermetically sealed cells can have adequate lifetime especially with ac operation. A dynamic-scattering reflective liquid-crystal display is usually backed with a specularly reflecting mirror. This means that the position of the observer relative to the display and the source of ambient light is critical if good contrast is to be obtained and is probably the reason that some people find such a display subjectively unsatisfactory.

To overcome these limitations, increasing emphasis is being placed on displays based on field effects in high-resistivity liquid crystals. One such effect is the action of an electric field in quenching the optical activity in a so-called "twisted nematic" cell [35]. In these devices, the electrode surfaces are treated so that the liquid crystal molecules are aligned with their long axes parallel to a specific direction in the plane of the surface. The preferred directions in the two electrode plates are made orthogonal, so that there is a progressive twist in the molecular axis across the cell which serves to rotate the plane of polarization of plane-polarized incident light. The resulting optical activity can be quenched by applying an electric field which, via the intrinsic molecular dielectric anisotropy, causes the molecules to stand on end, effectively destroying the twist. Displays based on this principle can have very good contrast and adequate viewing angle, although they are not particularly "bright" because at least half the incident light must be thrown away in the input polarizer. Good lifetime should be possible because no current need flow, although stability of the surfaces, which must be structured and coated with a surfactant to define the molecular orientation, is still a cause for concern. Another useful field effect is based on an electric field-induced transition from a scattering cholesteric phase to a clear quasi-nematic phase [36]. Such an effect can have a relatively sharp threshold and provide storage, obviously desirable attributes.

Displays that are potentially brighter and livelier than those provided by twisted nematic devices can be made using a pleochroic dye dissolved in a liquid crystal [37], or an electrophoretic colloidal dye suspension [33]. In these devices, the absorption properties of the dye can provide color contrast as well as luminance contrast, and result in a generally pleasing display. In the liquid-crystal dye cell, advantage is taken of the fact that when the dye molecules are aligned parallel to the direction of incoming light, little absorption takes place, whereas there is appreciable absorption for other orientations. The required dye molecule orientation can be induced by elastic interaction with the surrounding liquid crystal molecules whose alignment, in turn, can be determined by an applied electric field. By combining the guest-host interaction with the field-induced cholesteric–nematic transition, it appears possible to provide a display that not only has good brightness and contrast but also exhibits threshold and storage [38].

TABLE X

Type of Display	Threshold	Memory
Liquid crystal		
Dynamic scattering	weak	yes
Twisted nematic	better	no
Field-effect dye cell	good	yes
Electrophoretic	weak	yes

In electrophoretic displays one takes advantage of the fact that in a colloidal suspension of dye particles, individual particles are generally spontaneously electrically charged, and hence will move in an applied field. Thus if, for example, a thin cell is made up with white particles suspended in a black fluid between transparent electrodes, and a dc field is applied, the particles will "plate out" on one electrode and the display will appear white when viewed through that electrode. When the field is reversed, the particles will transfer to the opposite electrode and the display will appear dark. Displays based on this effect can have good brightness and contrast, with response times (~10 ms) adequately short for many applications.

As discussed earlier, the presence of a well-defined threshold and reliable memory is essential for economical matrix-addressing of all but the smallest displays. Table X summarizes the current status of liquid-crystal and electrophoretic displays as far as these features are concerned.

V. COMMENTARY

The preceding sections serve to reinforce what is already known to the reader, namely, there is an enormous range of available and potentially available solutions to display problems. Extrapolating from the past with a mix of technical insight, it appears highly likely that many of the potential displays will fail in the sense that they will never become a finished, successful product. There is no reason to expect 100-percent success. Indeed, we would be concerned with 100-percent success for it would imply too conservative an investment strategy in display research and development. This is hardly the case, however. Rather, we are concerned that too little justification is being required of display scientists and engineers, that the ratio of potential to risk is not being carefully evaluated, and that rather unconservative thinking is guiding the investment strategy. In the long run, this could only be harmful to the display community, to the general utility of displays, and to the financial health of the electronics industry.

One example might serve to illustrate the point. For the past 15 years, attention has been focused on the problem of achieving sufficient electroluminescent efficiency in III-V compounds to make viable an LED technology for efficient, reliable indicators and small displays. In the past few years, fond hopes have grown to gratifying reality with colors ranging from red through green, indicator lamps that are price competitive with miniature tungsten lamps while being more reliable, and numerics that have played an important role in the success of hand-held calculators. This success has been important not only to LED manufacturers and material suppliers but to the SIC industry.

The readers of this special issue are especially aware of the cost and difficulty of developing the material and processing technology to achieve this success. Even here there has been parallel effort: GaP with an indirect bandgap, GaAsP with a direct bandgap, and more recently GaPAs with an indirect bandgap.

In the midst of this success there are those who hold out the hope of blue LED's. It is true that it would be nice, and in some cases even useful, to round out the complement of colors. (There are those who argue that it is necessary to ultimately develop color TV displays of this type—an even more dubious goal!) But is it necessary or even desirable? Blue diodes would have to exhibit power efficiencies comparable to red ones to compete in brightness. This is not likely considering the trend in LED efficiencies as a function of wavelength. The eye has poor resolution for blue light. Given the cost of developing an entire new materials technology, the poor prospects for competing with a highly developed technology well down the learning curve, and the dubious advantage of providing one more color when there are already several, the quest for blue LED's surely does not represent a reasonable investment of research and development effort. Yet this investment is being made; might it not be more prudent instead to devote the same money to furthering the state of the art in the existing material systems where success might have more significant implications in the competition between LED's and other displays?

Given the spirit of these remarks, the reader may well perceive other examples. Perhaps he would rather not.

References

[1] M. F. Tompsett, W. J. Bertram, D. A. Sealer, and C. H. Sequin, "Charge coupling improves its image, challenging video camera tubes," *Electronics*, vol. 46, pp. 162–169, Jan. 18, 1973.

[2] E. I. Gordon, "Light addressed light valves," in *Conf. Rec., 1970 IEEE Conf. Display Devices* (New York, N. Y., Dec. 2–3, 1970), p. 29.

[3] P. Seats, "The cathode-ray tube—A review of current technology and future trends," *IEEE Trans. Electron Devices*, vol. ED-18, pp. 679–684, Sept. 1971.

[4] S. R. Hofstein, "Silicon-target storage tubes outdo direct-view types in versatility," *Electronics*, vol. 46, pp. 91–94, Feb. 15, 1973.

[5] A. Korpel, R. Adler, P. Desmares, and W. Watson, "A television display using acoustic deflection and modulation of coherent light," *Proc. IEEE*, vol. 54, pp. 1429–1437, Oct. 1966.

[6] Y. Yamada, M. Yamamoto, and S. Normura, "Large screen laser color TV projector," in *Dig. Tech. Papers, 6th Int. Quantum Electronics Conf.* (Sept. 1970), pp. 242–243.

[7] W. F. Goede, L. A. Jeffries, and J. E. Gunther, "A flat alphanumeric display tube," *IEEE Trans. Electron Devices*, vol. ED-18, pp. 692–697, Sept. 1971.

[8] G. E. Holz, "The primed gas discharge cell—A cost capability improvement for gas discharge matrix displays," in *1970 IDEA Symp., Dig. Papers* (May 1970), pp. 30–31.

[9] Y. S. Chen and H. Fukui, "A field interlaced real-time gas-discharge flat-panel display with gray scale," in *Conf. Rec., 1972 IEEE Conf. Display Devices* (New York, N. Y., Oct. 11–12, 1972), pp. 70–76.

[10] G. J. Chodil, M. C. DeJule, and J. Markin, "Good quality TV pictures using a gas-discharge panel," in *Conf. Rec., 1972 IEEE Conf. Display Devices* (New York, N. Y., Oct. 11–12, 1972), pp. 77–81.

[11] D. C. Krupka, Y. S. Chen, and H. Fukui, "On the use of phosphors excited by low-energy electrons in a gas-discharge flat-panel display," this issue, pp. 1025–1029.

[12] J. Forman, "Color in gas discharge displays," in *Proc. Soc. Inform. Displays*, vol. 13, pp. 14–20, May 1972. Also Z. van Gelder and M. M. M. P. Mattheij, "Principles and techniques in multicolor dc gas discharge displays," this issue, pp. 1019–1024.

[13] R. L. Johnson, D. L. Bitzer, and H. G. Slottow, "The device characteristics of the plasma display element," *IEEE Trans. Electron Devices*, vol. ED-18, pp. 642–649, Sept. 1971.

[14] W. D. Petty and G. H. Slottow, "Multiple states and variable intensity in a plasma display panel," *IEEE Trans. Electron Devices*, vol. ED-18, pp. 654–658, Sept. 1971.

[15] A. G. Fischer, "Design considerations for a future electroluminescent TV panel," *IEEE Trans. Electron Devices* (Corresp.), vol. ED-18, pp. 802–804, Sept. 1971.

[16] A. A. Bergh, and P. J. Dean, "Light-emitting diodes," *Proc. IEEE*, vol. 60, pp. 156–223, Feb. 1972.

[17] R. Solomon and D. DeFevere, "Efficiency shift in very high efficiency GaP(Zn, O) diodes," *Appl. Phys. Lett.*, vol. 21, pp. 257–260, Sept. 15, 1972.

[18] G. Marie, "Un nouveau dispositif de restitution d'images utilisant un effet electro-optique: Le tube TITUS," *Philips Res. Rep.*, vol. 22, pp. 110–132, Apr. 1967.

[19] P. M. Heyman, I. Gorog, and B. Faughnan, "High contrast thermal-erase cathodochromic sodalite storage-display tubes," *IEEE Trans. Electron Devices*, vol. ED-18, pp. 685–691, Sept. 1971.

[20] V. K. Zworykin and G. A. Morton, *Television, the Electronics of Image Transmission in Color and Monochrome.* New York: Wiley, 1954.

[21] R. J. Doyle and W. E. Glenn, "Lumatron: A high-resolution storage and projection display device," *IEEE Trans. Electron Devices*, vol. ED-18, pp. 739–747, Sept. 1971.

[22] J. R. Maldonado and A. H. Meitzler, "Strain-biased ferroelectric-photoconductor image storage and display devices," *Proc. IEEE*, vol. 59, pp. 368–382, Mar. 1971.

[23] J. R. Maldonado and D. B. Fraser, "PLZT ceramic display devices for slow-scan graphic projection displays," this issue, pp. 975–981.

[24] W. D. Smith and C. E. Land, "Scattering mode ferroelectric-photoconductor image storage and display devices," *Appl. Phys. Lett.*, vol. 20, pp. 169–171, Feb. 15, 1972.

[25] A. D. Jacobson, "Photoactivated liquid crystal light valves," in *1972 Int. Symp. Information Display, Dig. Tech. Papers*, pp. 70–71. Also J. D. Margerum, J. Nimoy, and S. Y. Wong, "Reversible ultraviolet imaging with liquid crystals," *Appl. Phys. Lett.*, vol. 17, pp. 51–53, July 15, 1970.

[26] M. Grenot, J. Pergrale, J. Doujois, and G. Marie, "New electro-optic light valve device for image storage and processing," *Appl. Phys. Lett.*, vol. 21, pp. 83–85, Aug. 1, 1972. Also G. Marie and J. Donjon, "Single-crystal ferroelectrics and their application in light-valve display devices," this issue, pp. 942–958.

[27] D. S. Oliver *et al.*, "Image storage and optical readout in a ZnS device," *Appl. Phys. Lett.*, vol. 17, pp. 416–418, Nov. 15, 1970.

[28] H. Melchior *et al.*, "Thermally addressed, electrically erased, high resolution, liquid crystal light valves," *Appl. Phys. Lett.*, vol. 21, pp. 392–394, Oct. 15, 1972. Also D. Maydan, "High-resolution graphic information," this issue, pp. 1007–1013.

[29] F. J. Kahn, "IR laser addressed thermooptic smectic liquid crystal storage displays," *App. Phys. Lett.*, vol. 22, pp. 111–113, Feb. 1, 1973.

[30] A. Sussman, "Electrooptic liquid crystal devices: Principles and applications," *IEEE Trans. Parts, Hybrids, Packag.*, vol. PHP-8, pp. 24–37, Dec. 1972.

[31] G. A. Castellion and D. P. Spitzer, "Solid state electrochromic mirror," U. S. Patent 3 713 719, Dec. 1970.

[32] A. M. Marks, "Electrooptical characteristics of dipole suspensions," *Appl. Opt.*, vol. 8, pp. 1397–1414, July 1969.

[33] I. Ota, J. Ohnishi, and M. Yoshiyama, "Electrophoretic display device," in *Conf. Rec., 1972 IEEE Conf. Display Devices* (New York, N. Y., Oct. 11–12, 1972), pp. 46–50.
——, "Electrophoretic image display (EPID) panel," this issue, pp. 832–836.

[34] G. H. Heilmeier, L. A. Zanoni, and L. A. Barton, "Dynamic scattering: A new electrooptic effect in certain classes of nematic liquid crystals," *Proc. IEEE*, vol. 56, pp. 1162–1171, July 1968.

[35] M. Schadt and W. Helfrick, "Voltage-dependent optical activity of a twisted nematic liquid crystal," *Appl. Phys. Lett.*, vol. 18, pp. 127–128, Feb. 15, 1971.

[36] J. J. Wysocki, J. Adams, and W. Haas, "Electric field-induced phase change in cholesteric liquid crystals," *Phys. Rev. Lett.*, vol. 20, pp. 1024–1025, May 1968. Also J. J. Wysocki *et al.*, "Cholesteric-nematic phase transition displays," in *1971 Int. Symp. Information Display, Dig. Tech. Papers* (May 1971), pp. 130–131.

[37] G. H. Heilmeier and L. A. Zanoni, "Guest-host interactions in nematic liquid crystals. A new electro-optic effect," *Appl. Phys. Lett.*, vol. 13, pp. 91–93, Aug. 1, 1968.

[38] D. L. White and G. A. Taylor, "A new liquid crystal absorptive mode display device," to be published.

Vapor Phase Epitaxial Materials for LED Applications

M. GEORGE CRAFORD AND W. O. GROVES

Abstract—The growth and performance of vapor phase epitaxial (VPE) III–V materials and devices for light-emitting diode (LED) and display applications are reviewed. Because of the commercial importance of the GaAsP ternary alloys and GaP, this paper is primarily concerned with these materials. With the addition of nitrogen doping high-performance red, yellow, and green devices can be fabricated using GaAsP and GaP. The VPE materials growth technology is highly developed, and, in conjunction with the Zn diffusion techniques used to form the p-n junctions, is compatible with planar processing procedures similar to those commonly used in the Si integrated-circuit industry. This processing technology results in reproducible and economical LED devices. In this paper the performance of VPE GaAsP devices is compared with the performance of other types of LED's, and economic considerations involved in the fabrication of LED devices are discussed.

I. Introduction

THE light-emitting diode (LED) industry has grown at an incredible rate over the past two years. It is estimated that at the present time more than three million LED devices per month are sold. This growth has been precipitated by an expanding number of LED applications including calculators, watches, and a host of other commercial and military devices. The use of LED's in the automotive industry, now only anticipated, should lead to further price reduction in LED's and to an even wider range of applications.

The growth of the LED industry has been possible because of the availability of a highly developed technology for the growth and fabrication of LED devices. This technology, namely the vapor phase epitaxial (VPE) growth of $GaAs_{1-x}P_x$ material and the formation of junctions by means of Zn diffusion in a planar process [1], [2], has permitted the fabrication of volume quantities of devices which meet reproducibly the performance specifications required for LED applications. With the introduction of nitrogen doping [3]–[6], high-performance VPE $GaAs_{1-x}P_x$ devices with colors ranging from red to green can be fabricated and further improvements in device performance are expected.

Other techniques for the fabrication of LED's have been investigated, in particular, an extensive research effort has been expended on liquid phase epitaxial (LPE) devices, predominantly red-emitting GaP doped with zinc and oxygen, GaP:Zn, O [7], [8]. Although GaP:Zn, O devices with high luminescent efficiencies have been developed in the laboratory [7], [8] the transition between the laboratory and the large volume requirements of the commercial marketplace has been difficult to accomplish.

In this paper the VPE crystal material and device technology and their use in LED and display applications are discussed. The performance of VPE devices is compared to those obtainable using the LPE technology. The discussion here is restricted to materials and devices that emit light in the visible portion of the spectrum.

Manuscript received January 26, 1973.
The authors are with Monsanto Commercial Products Company, St. Louis, Mo. 63166.

II. Review of III–V Material Systems for LED Applications

The III–V material systems with potential for LED applications have recently been reviewed by Archer [9]. These include the direct materials: $GaAs_{1-x}P_x$ ($x<0.49$) red, $Ga_{1-x}Al_xAs$ ($x<0.27$) red, $In_{1-x}Ga_xP$ ($x<0.70$) red to yellow, and $In_{1-x}Al_xP$ ($x<0.40$) red to green; and the indirect materials: GaP:Zn, O, red; GaP:N, yellow to green; $GaAs_{1-x}P_x$:N, red to yellow; $Ga_{1-x}Al_xP$, green; AlAs, yellow; and AlP, green. Another III–V material, GaN, emitting in the green [10], blue [11], or violet [12], depending on doping, might be included in this list. However, GaN, although a III–V compound, differs from the other III–V's in having the hexagonal wurtzite rather than cubic zinc blende structure, and shares the unfortunate characteristic of II–VI compounds in that it defies attempts to form the p-n junctions that are necessary for efficient minority carrier injection. Therefore, although promising photoluminescence results in GaN have been obtained, only relatively weak electroluminescence at high voltages characteristic of breakdown in insulating regions has been reported.

All of these materials have been or might be prepared by VPE methods. $GaAs_{1-x}P_x$ for red emitters and, to a lesser extent, $GaAs_{1-x}P_x$:N and GaP:N for yellow and green emitters, respectively, are presently in large-scale commercial production by VPE. Recent progress in these materials will be discussed in detail in this review. Recent improvements in the VPE process for preparing $In_{1-x}Ga_xP$ [13] have resulted in greatly improved $In_{1-x}Ga_xP$ electroluminescent (EL) diode performance; the best external quantum efficiency reported is 5.9×10^{-4} at 10 A/cm^2 and 6170 Å (unencapsulated diode), which is equivalent to the highest LPE $In_{1-x}Ga_xP$ value reported in the literature [9]. This result is also equivalent to the results obtained in the $GaAs_{1-x}P_x$:N alloy system for similar emission wavelengths [5]. However, due to a relatively large lattice mismatch and difference in the free energies of formation of the two end components ($x=0$ and $x=1$), the $In_{1-x}Ga_xP$ alloys are much more difficult to grow than $GaAs_{1-x}P_x$. Thus it is unlikely that this material system will achieve widespread use unless it can be demonstrated to have markedly superior performance or unless its technology is further developed and is simplified.

While the aluminum compounds have been prepared by VPE [14], the extreme reactivity of the aluminum chlorides requires use of protective liners in the quartz reaction systems and this, together with the high temperatures required, tends to increase the background impurity concentrations over those that can be achieved in the other materials. Although the $In_{1-x}Al_xP$ system offers potentially the highest luminous efficiency [9], the material synthesis problems would be the most severe. Problems associated with lattice mismatch between the two end components would be essentially the same as in the $In_{1-x}Ga_xP$ system, but there is an even wider disparity between the free energies of formation of the end components. The latter largely determines the normal

VPE growth temperatures, which are 650°C for InP [15] versus 1250°C for AlP [16]. This large energy difference suggests great difficulty in obtaining simultaneous deposition of the two components, at a single temperature.

Thus in principle, many III–V materials systems could be used for LED applications, but except for $GaAs_{1-x}P_x$ and GaP the material synthesis problems are formidable. Thus since $GaAs_{1-x}P_x$ and GaP already occupy such an important position in today's market, the remainder of this paper will be concerned almost exclusively with $GaAs_{1-x}P_x$ and GaP materials and devices.

III. Description of Basic VPE Growth Technique

A variety of techniques utilizing the vapor phase for epitaxial growth of III–V compounds have been described. Besides various vacuum evaporation [17], [18] and sputtering [19] techniques, a wide variety of chemical vapor transport [20] or chemical vapor deposition methods including the early closed-tube [21], [22], close-spaced [23], and open-tube flow methods [1], [24] have been employed. A variant of the open-tube flow method, which may become more important for the growth of the more difficult aluminum and nitride compounds and alloys, involves the pyrolitic decomposition of group III metal organics in the presence of group V hydrides [25]–[28].

Because of its flexibility and the degree of control that can be achieved over the growth conditions determining composition, purity, and dopant concentrations, as well as amenability to scale-up into a manufacturing process [1], [24], the open-tube flow method has come into the widest usage.

A number of chemical reactions may be used to achieve VPE growth of III–V materials. In general, all of these include: 1) generation of volatile group III metal species by reaction of either the metal or one of its compounds with a transport agent; 2) introduction of the group V component; 3) addition of a dopant species, if required; and 4) passage of the reactant gas mixture over suitably prepared single-crystal substrates in a temperature region where single-crystal growth will occur.

The volatile group III metal species are preferable in the form of the subhalides because the higher metal-to-halide ratios and higher deposition temperatures lead to more efficient deposition and better crystalline structure than can be achieved by reduction of the trihalides [29]. Water vapor, which forms the volatile Ga_2O, has been used as the transport agent in growing GaAs [30], $GaAs_{1-x}P_x$, and GaP [31] successfully, but leads to the incorporation of oxygen which ordinarily is not desired in the crystal. Furthermore, this growth technique requires higher temperatures and therefore perhaps higher levels of background contamination. Of the halides, the most effort has gone into development of chloride systems, in which either the group V trichlorides or anhydrous HCl are used as the primary transport reactants. There are a number of advantages to using HCl rather than the group V trichlorides: 1) the equipment is simple, only gas flow control being required; 2) concentrations of the group III and group V components may be varied independently, simplifying variation and control of group III–V ratios; 3) in mixed group V alloy systems, since the group V elements need not contact the source, their ratio can be independently varied and controlled without reference to the previous history of the source or without resorting to separate sources [1], [32].

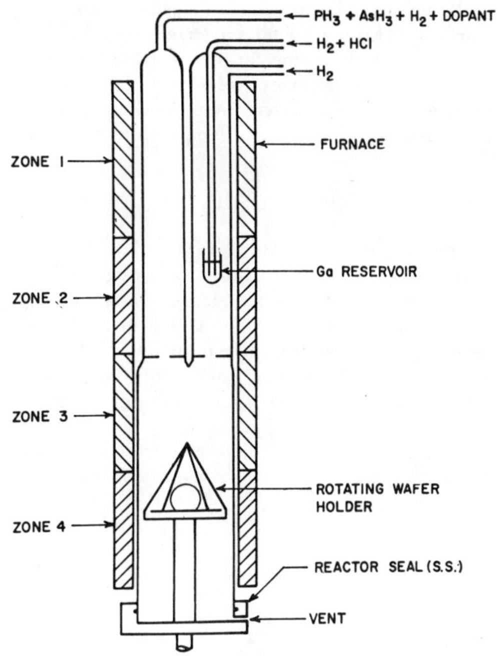

Fig. 1. Schematic diagram of vertical multiwafer reactor for vapor phase epitaxial deposition of $GaAs_{1-x}P_x$ [1].

In addition, to date, higher purities have been achieved in GaP [33] from the Ga–HCl–PH_3 system than from Ga–PCl_3 systems [34] although this is not the case for the analogous GaAs system [35].

While the use of compounds or homogenized mixtures of compounds as the group III metal source offers advantages in some cases, in general, for open-tube flow systems, operating convenience, as well as the higher purity of group III metals available in elemental form, dictates use of the latter. With HCl transport, the group V hydrides, AsH_3, PH_3, and NH_3, are the most convenient sources of the group V component [14]. Arsenic and phosphorous in elemental form, located in controlled temperature zones in the furnace have also been used [36], [37], but in this form are less convenient, are of generally lower purity, and are more expensive than the hydrides. Various dopants are most conveniently added as gaseous or volatile compounds to one of the reactant gas streams or evaporated in the elemental form from a separately controlled temperature zone in the furnace.

For the usual temperatures and in the presence of a large excess of H_2 the III–V deposition reaction may be written in the form,

$$MCl(g) + 1/zN_z(g) + 1/2H_2(g) \rightarrow MN(s) + HCl(g) \quad (1)$$

where M represents a group III metal, Ga, In, or Al; N represents As or P; and z is the number of atoms in the molecule. Since the enthalpy of this reaction is negative, it will proceed to the right as the temperature is lowered. Hence the substrates are located in a zone at lower temperature than that at which the reactant gases are initially brought together.

The most widely used VPE growth method is illustrated in Fig. 1 which shows a schematic diagram of a multiwafer reactor of the type used for the commercial production of

$GaAs_{1-x}P_x$ epitaxial layers. As has been demonstrated by Tietjen *et al.* [14] using a similar system but of different geometry, this type of reactor can be used, with appropriate modification, for a wide range of III–V material systems. HCl, diluted with hydrogen, is metered into the Ga reservoir in one arm of the reactor in zone 2 at a temperature of about 775–800°C. Here the reaction

$$Ga(l) + HCl(g) \rightleftarrows GaCl(g) + 1/2H_2(g) \quad (2)$$

takes place. The GaCl–HCl–H_2 mixture is diluted by an additional stream of H_2 and swept into the mixing zone, zone 3, where it combines with a third stream containing arsenic, phosphorus, and dopant from the other arm of the reactor. The latter is formed by a mixture of arsine, phosphine, and the dopant gas with more hydrogen. Group VI dopants, S, Se, or Te are added by metering the dilute mixtures of H_2S, H_2Se, or $(C_2H_5)_2Te$, respectively. Doping with the isoelectronic impurity N can be accomplished by adding NH_3 to the dopant gas mixture. To introduce a p-n junction into the crystal, the group VI dopant flow may be interrupted and a $(C_2H_5)_2Zn$ mixture metered in through a different line.

With substrate temperatures of 800 to 825°C, growth rates of ~1 μm/min are typically obtained for GaAs. With increasing phosphorus content, the growth rate generally decreases to ~0.4 μm/min for $GaAs_{0.6}P_{0.4}$ and to ~0.2 μm/min for GaP. In a typical run, crystal growth is initiated as 100-percent GaAs or GaP, according to the substrate used. The AsH_3 and PH_3 flow rates are then gradually changed to produce a compositionally graded alloy layer before the proper ratio in the gas phase is reached to produce the desired solid alloy composition. This serves to minimize the disorder produced by the lattice mismatch between substrate and layer.

As a typical example, the reactor shown in Fig. 1 was designed to accommodate eight 20-mm diameter substrates [1] or a total of ~4 in². Modifications in the reactor system, coupled with the development of larger area GaAs substrates, have increased by an order of magnitude the capacity of modern VPE reactors used to produce commercial quantities of $GaAs_{1-x}P_x$, the demand for which is now more than one million square inches annually [38].

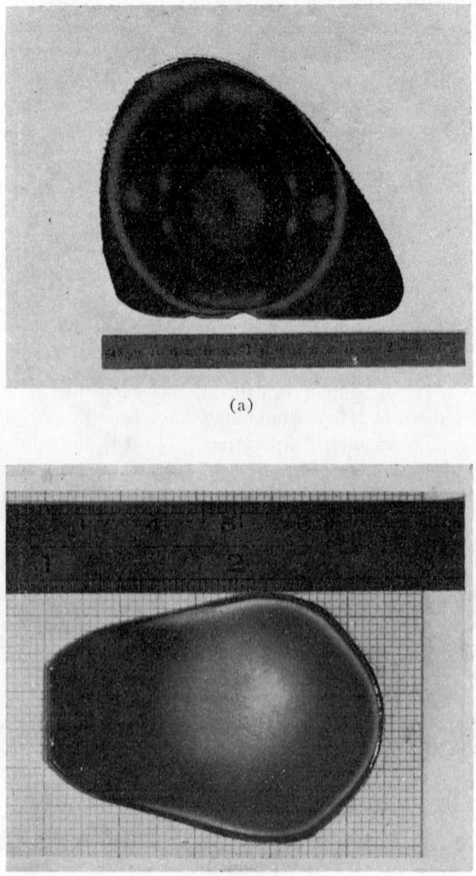

(a)

(b)

Fig. 2. $GaAs_{1-x}P_x$ epitaxial wafers. (a) On GaAs substrate, area ~3 in² (wafer provided by J. Burd). The reflection of the camera can be seen in the surface of the wafer. (b) On GaP substrate, area ~1 in² (wafer provided by D. Finn).

IV. Structural and Electrical Properties

A. *Wafer Size and Surface Properties*

The useful area of VPE III–V single crystal wafers is limited by the size of available substrates. While some work has been done on other substrate materials, Ge for $GaAs_{1-x}P_x$ [39] and Si for GaP [40], it has been difficult or impossible to achieve the required structural or electrical characteristics in the grown layers. Meanwhile, continued progress in the development of ever larger GaAs and GaP single crystal ingots has reduced the pressure to find alternate materials.

Wafers which illustrate the size and surface quality of commercial $GaAs_{1-x}P_x$ wafers grown on GaAs and GaP substrates, respectively, are shown in Fig. 2(a) and (b). The diameter of the GaAs wafer is greater than 2 in and the surface area is greater than 3 in². The surface quality is satisfactory for the use of planar junction processing similar to that used in conventional silicon technology. The VPE layer thickness uniformity is typically constant to within ~10 percent and the alloy composition is constant to within ≤1 percent across the surface of the wafers [37].

Currently available GaP wafers average ~1 inch in diameter or ~0.8 inch² in area. The thickness variation across a typical vapor epitaxial GaP, or high phosphorus content $GaAs_{1-x}P_x$ alloy wafer, as shown in Fig. 2(b), is nonlinear. A concave profile with about ±10-percent thickness variation from center to edge is superimposed on the 10-percent edge-to-edge variation normally seen in 40-percent $GaAs_{1-x}P_x$. This does not include an edge ridge, which occupies ~2 mm at the outer edge, depending on layer thickness.

The cross-hatch pattern which is visible on the surface of the wafers shown in Fig. 2(a) and (b) is typical of $GaAs_{1-x}P_x$ wafers grown by VPE on near ⟨100⟩ oriented substrates and is related to the array of misfit dislocations formed during growth of the initial compositionally graded portion of the epitaxial layer [41]–[44]. Similar patterns are also seen on high-quality $Ga_xIn_{1-x}P$ wafers grown on GaAs or GaP substrates with graded intermediate layers [13].

A cross sectional view of a typical VPE structure is shown in Fig. 3. The graded region, in which the alloy composition is graded from $x=0$ or $x=1.0$, depending on the substrate material, to the final desired alloy composition, has a thickness of 25 to 40 μm or a gradient of ~1 percent/μm. This graded region is necessary in order to reduce the density of dislocations and other crystal imperfections due to lattice mismatch between the $GaAs_{1-x}P_x$ layer and the substrate

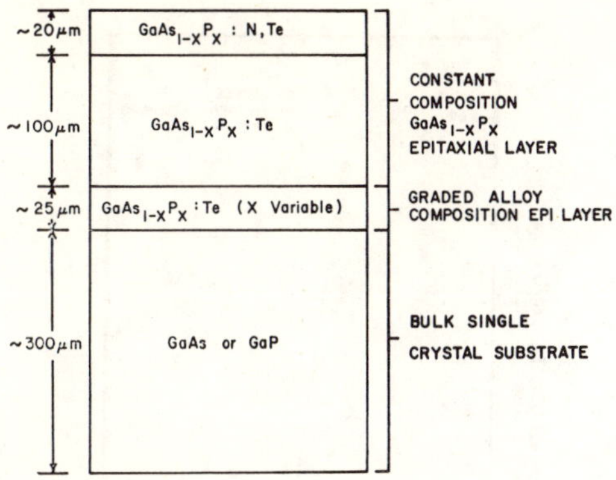

Fig. 3. Cross-sectional view of $GaAs_{1-x}P_x$ vapor phase epitaxial structure. The substrate is chosen to be GaAs or GaP depending upon which best matches the desired alloy composition. The nitrogen is introduced during the growth of the final ~20 μm by adding NH_3 to the dopant gas stream.

[41]. The dislocation density in the grown layer is believed to be proportional to the maximum composition gradient in the graded region [44]. If the composition gradient is too steep, the resulting high dislocation density in the alloy layer reduces its luminescence efficiency [41], [43], [44]. However, increasing the thickness of the graded layer is normally accompanied by concave bowing of the wafers unless certain.

Stringfellow [47] has shown that in LPE growth of $Ga_xIn_{1-x}P$ on GaAs substrates, the constraint of the GaAs lattice causes nonequilibrium growth of the alloy composition, $x=0.51$, matching the GaAs lattice constant, from a wide range of Ga–In–P liquid compositions. The extent to which a similar mechanism might operate in the vapor systems is difficult to assess but may contribute to the difficulty of growing alloy systems such as $Ga_xIn_{1-x}P$ with large lattice mismatch.

B. Background Impurities and Stoichiometric Defects

The crystal purity is a critical parameter in the growth of these materials since significant trace concentrations of certain impurities can result in poor device performance.

The background impurity concentrations for $GaAs_{1-x}P_x$ grown in a system such as described in Section III can range from below 10^{15} net donors per cubic centimeter for GaAs and $GaAs_{1-x}P_x$ ($x \leq 0.4$) up to $\sim 5\times10^{16}$ net donors per cubic centimeter for GaP [48]. Often, due to system or reactant feed contamination or changes in operating conditions, background impurity concentration can range much higher. The impurities in a typical production $GaAs_{1-x}P_x$ sample as determined by spark source mass spectrometry [49] are shown in Table I. The conversion factor for parts per million atomic into atoms per cubic centimeter is $\sim 5\times10^{16}$. At these low impurity levels it is difficult to correlate electrical activity with trace impurity concentrations. In GaP where the background impurity concentration is higher, good correlations with Si contamination have been obtained [37], [48].

Weiner [50] has treated the problem of Si contamination in open-flow systems from an equilibrium thermodynamic viewpoint, and shown that reaction of HCl with quartz can lead to significant contamination of the gas stream with volatile Si compounds and water vapor. On the other hand, since a liquid-Ga source will remove most of the Si from the vapor which contacts it, as well as reducing the free HCl to low concentrations, synthesis systems, such as in the production process previously described, should appreciably reduce Si contamination.

TABLE I
IMPURITIES IN A TYPICAL $GaAs_{1-x}P_x$ SAMPLE BY MASS SPECTROMETRY [49]

Elements	ppm Atomic
C	1.1
N	0.2
O	0.9
Al	0.04
Si	0.04
S	0.05
Cl	0.06
K	0.03
Ca	0.03
Cr	<0.01
Fe	<0.01
Cu	interference
Te (dopant)	1.0

The low-temperature photoluminescence (PL) spectra of high-purity GaP material grown on the ⟨100⟩ orientation, in contrast to results previously published, for samples slightly contaminated with S [33], contain only a trace of the strong no-phonon S line, indicating an S content well below 10^{15} atom/cm^3 [51]. However, the PL spectra of these, as well as the S contaminated samples, all contain prominent lines now attributed to Si [52]. The absence of S also markedly alters the 77 K EL spectrum [51]. Lines involving S are absent while two bands at 6020 and 7100 Å are observed, undoubtedly due to Si and probably related to the broad orange band at 6300 Å which Lorentz and Pilkuhn [53] have attributed to Si, their samples being much more heavily doped.

Another characteristic of the PL and EL spectra of these high-purity GaP samples is the absence of the nitrogen line commonly found as a contaminant in GaP samples [33]. The intensity of the nitrogen line in PL is less than 10^{-3} of the intensity of the no-phonon sulfur line in a sulfur contaminated sample containing 1.8×10^{16} total donors. Thus in contrast to LPE GaP, virtually nitrogen-free GaP can be prepared by vapor phase epitaxy unless N doping is deliberately introduced.

These results demonstrate the purity that can be achieved in an all quartz reactor using the H_2–HCl–Ga–PH_3 system. In the same system GaAs has been produced with $<10^{15}$ net donors per cubic centimeter and mobilities of up to 8000 and 90 000 at 300 K and 77 K, respectively. The transport properties of the GaP and $GaAs_{1-x}P_x$ alloys produced in this system will be discussed in the next section.

In addition to minimizing undesirable background impurities, the material growth conditions must be adjusted to minimize native defects. Stewart [54] has found that, under certain growth conditions, varying the group III–V ratio and Se dopant concentrations has a pronounced effect on the intensity of an IR band at ~1.3 eV. The presence of such a competing recombination mechanism, of course, reduces the luminescence efficiency of the desired bandedge radiation. The increase in the relative intensity of the IR emission with decreasing group III–V ratio and increasing Se concentration parallels the expected increase in Ga vacancy concentration and association of Ga vacancies with Se. Stewart therefore suggests that the IR radiation is due to formation of the neutral ($V_{Ga}+3$ Se) complex first postulated by Vieland and Kudman [55]. This complex is the stoichiometric equivalent of a Ga_2Se_3 molecule.

Additional evidence for the generation of native defects in III–V crystals is found in the analysis of temperature dependence of electrical properties of donor doped crystals [33], [56]. Also, carrier lifetime measurements indicate a residual defect density 10^{15}–$10^{17}/cm^3$ [57]. Increasing the dopant concentration generally results in an increase in acceptor concentration as well. The self-compensation mechanism, by which native acceptor defects (cationic vacancies) are automatically generated on addition of donor impurities, is well established for II–VI compounds [58]. A similar mechanism could account for the preceding phenomenon. Stewart's [54] work has shown that this self-compensation might be offset by adjusting the reactant gas stoichiometry.

C. *Electrical Properties*

Measurement of the electrical transport characteristics, carrier concentration, and electron or hole mobility of electroluminescent III–V materials is important because, as a rule, a good correlation is found between these properties and luminescent efficiency [2], [33].

Generally, the III–V alloy systems of interest are direct energy bandgap semiconductors for alloy compositions $x<x_c$, and indirect for alloy compositions $x>x_c$, where x_c is the composition of the "crossover," the point at which the direct Γ and indirect X minima are equal in energy. At compositions well below the crossover x_c the mobilities are high since the energy bandgap is direct and the conduction band effective mass is relatively small, similar to GaAs or InP. At compositions, $x>x_c$, the mobilities are low since the indirect conduction band effective mass is large, as in GaP or AlAs. As the crossover composition x_c is approached, the separation between the direct and indirect conduction band minima becomes comparable to thermal kinetic energy kT. As a result, and because of the higher density of states in the indirect minimum, a rapidly increasing number of electrons is transferred to the indirect minima. With two types of carriers involved in the conduction process, the mobility can be written [59]:

$$\mu_H = \frac{n_1\mu_1^2 + n_2\mu_2^2}{n_1\mu_1 + n_2\mu_2} = \mu_1 \frac{1 + (n_2/n_1)(\mu_2/\mu_1)^2}{1 + (n_2/n_1)(\mu_2/\mu_1)} \qquad (3)$$

where n_1 and n_2 are the carrier densities and μ_1 and μ_2 are the mobilities in the direct and indirect conduction bands, respectively. Assuming a Boltzmann distribution of carriers in the two bands, with reasonable values for the density of states and mobility ratios, it can be seen that a sharp drop in mobility will occur at values of x well below the crossover.

The sensitivity of electron mobility to alloy composition in the range immediately below the crossover, as shown in Fig. 4, obviates the utility of direct mobility measurements in assessing material quality of $GaAs_{1-x}P_x$ with $x \sim 0.40$, for example. However, such measurements for samples with compositions well removed from the crossover range can be used to evaluate the synthesis system. This is particularly true for GaP, itself.

The electron mobility in GaP at room temperature is dominated by lattice scattering so that significant effects of impurity or other scattering mechanisms can usually be seen only at low temperatures. The Hall mobility as a function of temperature for several samples of high-purity GaP [33] grown by a VPE method is shown in Fig. 5. These mobility values, as high as 2730 $cm^2/V \cdot s$ at 77 K, are the highest yet reported for GaP.

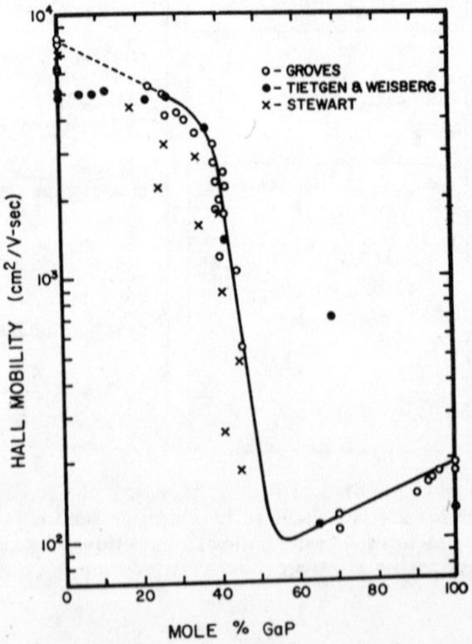

Fig. 4. Hall mobility versus alloy composition for $GaAs_{1-x}P_x$ with carrier concentration $<10^{16}/cm^3$. From [60]–[62].

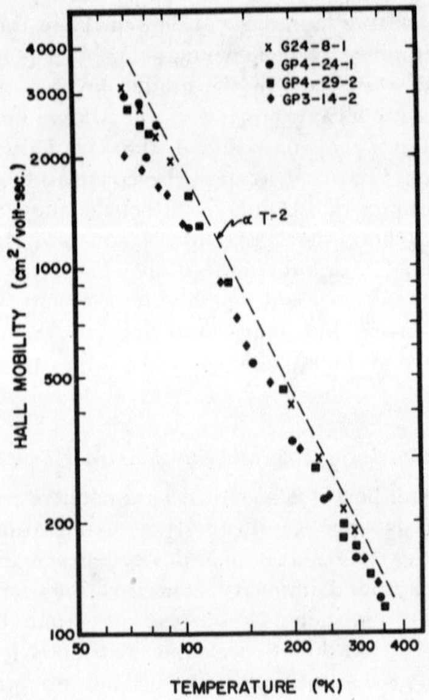

Fig. 5. Hall mobility as a function of temperature for n-type GaP [33].

Electrical characteristics have been studied in any detail only for the $GaAs_{1-x}P_x$ system. The Hall mobility as a function of alloy composition for samples with low carrier concentration, as measured by Tietjen and Weisberg [60], Stewart (cited by Lees) [61], and Groves [62] is shown in Fig. 4. The sharp drop in mobility near $x=0.49$, i.e., near the crossover x_c is expected on the basis of (3). However, the decrease in mobility between $x=0$ and $x=0.3$ is considerably larger than that expected due to the increasing effective mass [61], and the increase between $x=0.5$ and $x=1.0$ is totally unaccounted for on the basis of the simple two-band model. Tietjen and Weisberg [60] have estimated that alloy scattering should contribute less than ~10-percent reduction in mobility between $x=0$ and $x=0.3$ based on the relatively small values of the effective mass and, similarly, the reduction between

$x = 1.0$ and $x = 0.5$ should be less than ~ 50 percent [6]. Clustering of As and P atoms, not taken into account by the basic alloy scattering theory, may account for the additional variation in mobility [60] and is consistent with the carrier lifetime data of Zwicker *et al.* [57]. It would be interesting to compare similar data, when available, for good samples in the $In_{1-x}Ga_xP$ system where such clustering is expected to be even more prominent [63].

In $GaAs_{1-x}P_x$ alloys with x fixed at a given value below the crossover $x_c = 0.49$ mobilities increase monotonically with decreasing impurity concentration well down into the $10^{14}/cm^3$ range, with no indication of attaining a lattice scattering or other mobility limit [62]. The slopes of the family of curves of mobility versus log carrier density for each value of x decrease regularly with increasing x. The decrease in mobility with increasing carrier concentration appears stronger than expected for simple ionized impurity scattering. A mobility–donor concentration relationship recently reported [64] for $GaAs_{1-x}P_x$ with $x = 0.38$ with significantly higher mobilities at 10^{17} donor/cm^3 is believed to be in error. There is no indication that the graded layer, essential to grow crystals of the quality reported [65], was removed before making the electrical measurements.

The importance of removing the region of graded composition from an epitaxial layer before making valid Hall and resistivity measurements cannot be overlooked. Significant errors can result if the graded layer is included. This is because the mobility of the graded layer, particularly near the GaAs substrate is high, whereas the mobility of the constant composition region, especially for $x \gtrsim 0.4$, is low. The average measured mobility is weighted in favor of the higher mobility by the relation

$$\mu_H = \frac{n_1 \mu_1^2 t_1 + n_2 \mu_2^2 t_2}{n_1 \mu_1 t_1 + n_2 \mu_2 t_2} \qquad (4)$$

similar to (3), but where t refers to the thickness of each layer and the subscripts apply to the constant composition and graded layers, respectively. For thick layers, $n_1 \gg n_2$, with small x, $\mu_1 \sim \mu_2$, the error is small, but for thin layers, with large x, $\mu_1 \ll \mu_2$, the error can be quite large, and could easily account for the anomalously high mobilities reported in [64].

Another factor that must be considered in specifying donor concentrations of alloys with compositions near the crossover, is that due to the two carrier conduction mechanism, Hall measurements do not give the true total carrier concentration, $n_1 + n_2$. From the simple two-band model, the Hall concentration could be in error by as much as a factor of 8 near the crossover where $(n_2/n_1) = (\mu_1/\mu_2)$. Capacitance–voltage or avalanche breakdown measurements, which yield true total carrier concentrations, actually indicate a discrepancy of a factor of 3 near the crossover [66]. This indicates the importance of indicating the measurement method when specifying donor concentrations of $GaAs_{1-x}P_x$ for red-light emitters.

V. Electroluminescent (EL) Properties

A. Radiative Recombination Processes in $GaAs_{1-x}P_x$

The processes responsible for radiative recombination in $GaAs_{1-x}P_x$ have been discussed in several publications [2]–[6], and will only be briefly summarized here. In order to resolve the processes responsible for radiative recombination it has been useful to study the EL at 77 K as well as 300 K since at the lower temperature the resolution is better and individual transitions can be identified. The peak emission energies as a function of alloy composition for $GaAs_{1-x}P_x$ without and with nitrogen doping are shown in Figs. 6 and 7, respectively. In the direct bandgap region ($x < 0.46$ at 77 K) there are two peaks visible in nitrogen-free diodes [2]. The higher energy peak, which corresponds closely to the direct energy gap E_Γ is attributed to band-to-band recombination [2]. The second peak approximately 0.030 eV lower in energy is attributed to recombination from the conduction band to neutral zinc acceptor states [2]. In the indirect bandgap region ($x > 0.46$) three peaks are visible, corresponding to free exciton emission: E_{gx}, longitudinal acoustic (LA) phonon assisted free exciton emission $E_{gx} - LA$, and shallow donor–acceptor pair recombination [2], [4].

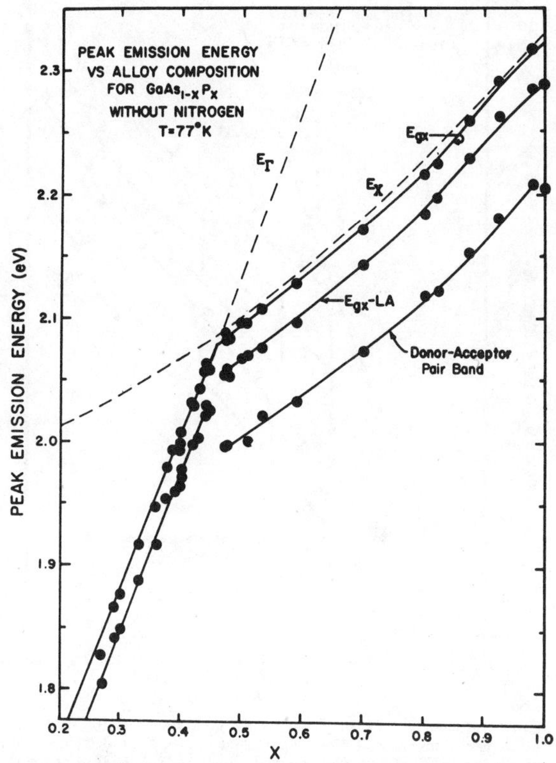

Fig. 6. Peak emission energy versus alloy composition for $GaAs_{1-x}P_x$ diodes without nitrogen at 77 K [4]. Many of the data for $x < 0.5$ were taken from [2]. The position of E_Γ and E_X was determined as described in [4]. The diode current is 5 mA (5 A/cm^2).

When the isoelectronic impurity nitrogen is added to the crystal, recombination associated with the nitrogen dominates the radiation [3]–[5]. For alloy compositions close to GaP three separate transitions can be observed. The highest energy line is the A-line recombination which is due to bound exciton recombination at isolated N-atoms [4]. The other two lower energy transitions are NN_3 and NN_1 transitions, which are attributed to recombination at centers consisting of nitrogen atoms at third nearest neighbor and nearest neighbor sites, respectively [4]. For alloy compositions $x < 0.8$, the A-line is no longer observed, and the NN_1 and NN_3 transitions cannot be separately resolved, so the recombination is simply labeled NN recombination in Fig. 7. These data shown in Fig. 7 were taken on samples in which the NH_3 flow rate in the crystal synthesis was held constant for all alloy composi-

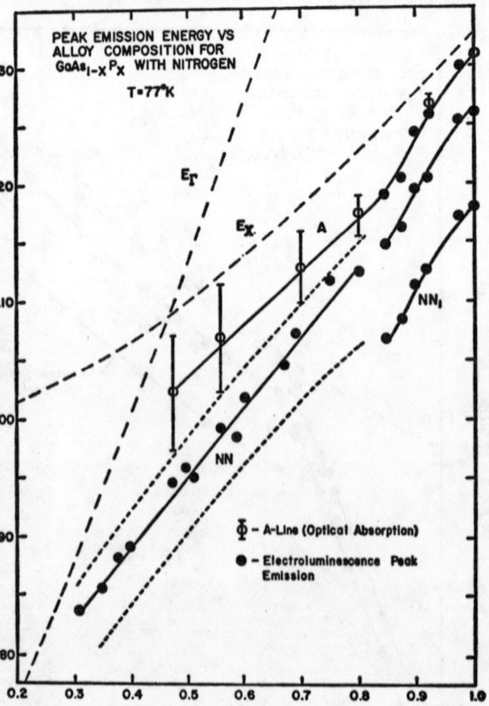

Fig. 7. Peak emission energy versus alloy composition for $GaAs_{1-x}P_x$ diodes with nitrogen at 77 K [4]. The curves for E_Γ and E_X are taken from Fig. 6. The peak position and width at half maximum for the optical absorption data are shown. The dashed lines indicate the half-power points for the NN pair band luminescence. The diode current is 5 mA (5 A/cm²).

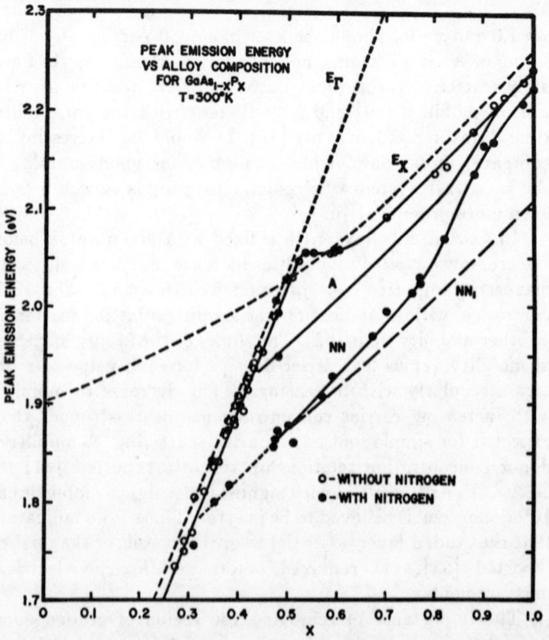

Fig. 8. Peak emission energy versus alloy composition for $GaAs_{1-x}P_x$ diodes with and without nitrogen doping at 300 K. The position of E_Γ was taken from [15]. The position of E_X and the A-line have been adjusted from the 77 K data as described in [4]. The estimated position of the NN_1 pair band is shown. The diode current is 5 mA (5 A/cm²).

tions. The flow conditions used resulted in an N concentration of $\sim 10^{19}/cm^3$ in GaP, but the concentration in the alloys has not been determined [4]. It is possible by reducing the N concentration to resolve the A-line transition in EL for alloy compositions $x<0.8$. Both the A-line and NN_3 transitions have been observed by means of optically pumping GaAsP:N platelets for alloy compositions of $x<0.50$ [67].

As the temperature is increased to 300 K, the radiative transitions closer to the conduction band edge generally dominate the recombination [4]. The peak emission energy as a function of alloy composition at 300 K is shown in Fig. 8. For direct bandgap alloys, without N doping, the band-to-band transition dominates over the band-to-acceptor transition because at 300 K an appreciable number of electrons occupy the Zn levels, giving rise to holes in the valence band, which results in an increased probability for band-to-band transitions [2].

For indirect bandgap alloys, at 300 K without N doping, the free exciton and phonon assisted free exciton recombinations dominate over the shallow donor–acceptor pair recombination since the recombination lifetime is long and the carriers tend to be thermally excited out of the pair states. The exciton processes have shorter recombination lifetimes and therefore can compete more effectively with the nonradiative processes [2]–[4]. In the case of nitrogen-doped diodes with $x<0.8$ the NN pair band continues to dominate the recombination at 300 K as it does at 77 K. For alloy compositions $x>0.8$, however, the A-line and its phonon replicas dominate instead of the NN_3 and NN_1 pair bands [4]. An exception to this is the case of GaP with nitrogen concentrations in excess of $10^{20}/cm^3$, which have extremely high pair concentrations [68], [69]. In devices of this type the NN recombination continues to dominate the recombination at 300 K as it does at 77 K [68], [69]. Thus these devices behave similarly to the GaAsP:N alloys with $x<0.8$. The change in behavior at $x \sim 0.8$ is not well understood, and may simply be due to a change in the nitrogen concentration which is incorporated as a function of alloy composition. It might be possible by increasing the N concentration in the region from $x=0.8$ to $x=1.0$, to force the NN pair band to dominate for all alloy compositions. In this case the peak recombination would approximately follow the dashed curve shown in Fig. 8. Alternatively, by sufficiently reducing the nitrogen concentration, it might be possible to allow the A-line to dominate for all alloy compositions.

B. The EL Performance of VPE $GaAs_{1-x}P_x$ and GaP

In this section the EL performance of $GaAs_{1-x}P_x$ LED's at 300 K will be reviewed. The emission spectra as a function of alloy composition, external quantum efficiency, brightness, and luminescent power efficiency will be discussed. The GaAsP devices used for these measurements were full chip emitters with areas of 10^{-3} cm² [4]. Except where noted otherwise, the current density was 10 A/cm² (10 mA through the device) which is a typical operating current density. The devices were mounted on TO-18 headers with the entire n-type side alloyed to the header and a wire bond was attached to the p-type diffused side. Except where noted otherwise, the diodes were not encapsulated in epoxy. This package was chosen for convenience and gives a good relative measure of $GaAs_{1-x}P_x$ with different alloy compositions. The diode

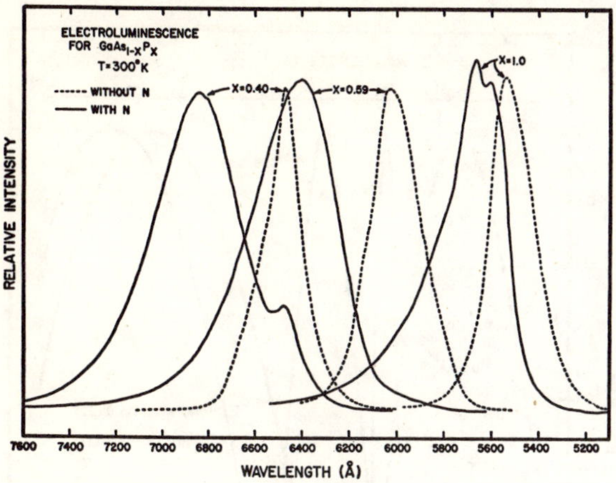

Fig. 9. Emission spectra from $GaAs_{1-x}P_x$ diodes with and without nitrogen doping [5].

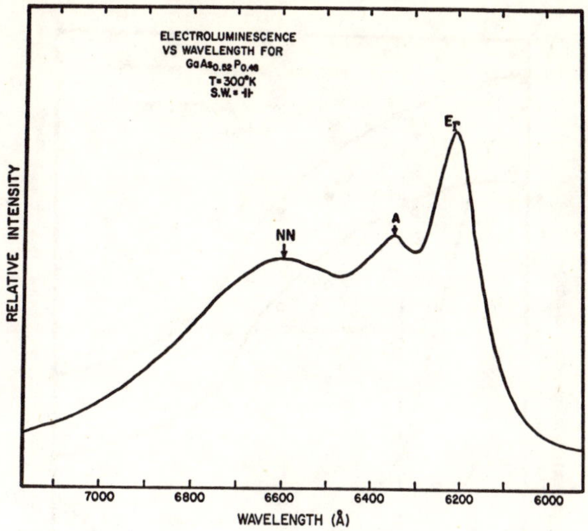

Fig. 10. Electroluminescence for a nitrogen-doped $GaAs_{0.52}P_{0.48}$ sample at 300 K [4]. This sample has $\sim 5x$ less than the other diodes studied and as a result the A-line is visible in addition to the E_Γ and NN emission. The diode current is 5 mA (5 A/cm²).

package is not optimized for high-efficiency performance since the alloy n-type contact is quite absorptive. It should be possible to obtain a marked increase (possibly $>2x$) in efficiency by resorting to a mesa structure with a polished n-type surface mounted on glass [70a].

The brightness values reported here for GaAsP diodes are measured values obtained using a calibrated photo research spotmeter with a spot diameter of 0.008 in. The efficiency measurements were carried out using a silicon solar cell as described elsewhere [2] and are known to be conservative by approximately 30 percent when compared to an integrating sphere. The brightness measurements as well as the efficiency measurements are average values for a batch of 10–20 diodes.

The emission spectra for several GaAsP alloy compositions with and without nitrogen doping are shown in Fig. 9. It can be seen that for a given composition the peak emission energy occurs at a shorter wavelength, or high energy, without nitrogen and that the full width at half height of the emission is narrower without the nitrogen. This behavior is of course expected from the discussion in the preceding section since the nitrogen emission occurs from a broad band of NN pair states located beneath the conduction band edge. In the case of the nitrogen-doped samples with $x=0.4$ there are two peaks visible corresponding to the direct bandgap Γ transitions and the NN pair line [3], [4]. The nitrogen-free sample with $x=0.4$, which has only the Γ emission, is equivalent to the $GaAs_{1-x}P_x$ diodes commercially available with a peak emission wavelength of approximately 6500 Å. The reason that, for $x=0.4$, both the NN and Γ emission are observed is that in the direct bandgap region the Γ emission is extremely efficient and competes effectively with the NN pair transitions [3], [4]. This is, of course, not the case in the indirect bandgap region where the band-to-band transitions are forbidden by momentum conservation requirements. All of the emission spectra shown in Fig. 9 came from samples with heavy nitrogen concentrations. When the nitrogen concentration was reduced, the A-line could be resolved in addition to the NN pair line as shown in Fig. 10. For this sample the NH_3 dopant gas flow was reduced by approximately $5x$.

The peak emission energy as a function of alloy composition is shown in Fig. 8. The dashed lines show the position of the direct and indirect conduction band edges and the position of the A-line as well as the NN pair band which is a combination of the NN_1 and NN_3 pair bands. As described in the previous section, the dashed line is shown which connects the position of the NN pair band in the $GaAs_{1-x}P_x$ alloys with $x<0.8$, with the pair emission band which can be observed in GaP with extremely heavy nitrogen doping [68], [69]. For N doping levels of $>10^{20}$ the emission in GaP is yellow instead of the yellow-green A-line and phonon assisted A-line emission characteristic of nitrogen concentrations in the vicinity of $\sim 10^{19}/cm^3$ [68], [69]. Emission spectra for VPE GaP diodes with a variety of nitrogen concentrations are shown in Fig. 11 [69]. As can be seen, as the nitrogen concentration is increased the peak emission wavelength increases from 5530 Å, where free exciton recombination dominates, to 5870 Å, where NN pair recombination accounts for the majority of the recombination [68], [69]. These yellow-emitting GaP diodes

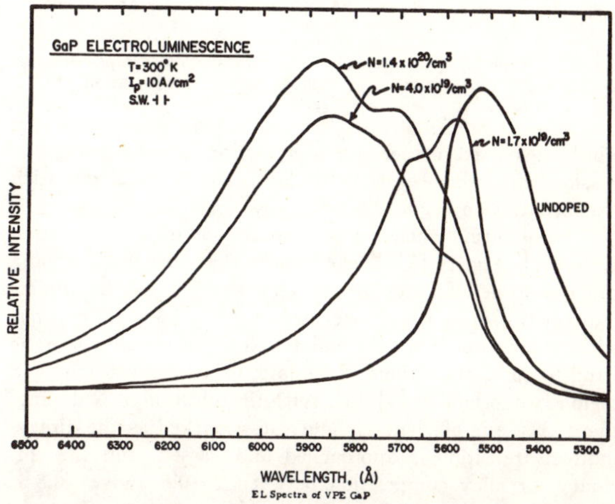

Fig. 11. Electroluminescence for VPE GaP diodes with different nitrogen concentrations. The emission from the undoped sample has a deep green color, whereas the sample with N concentration $=1.4\times10^{20}$ /cm³ has a yellow color. The N concentration was measured by means of optical absorption as described in [70].

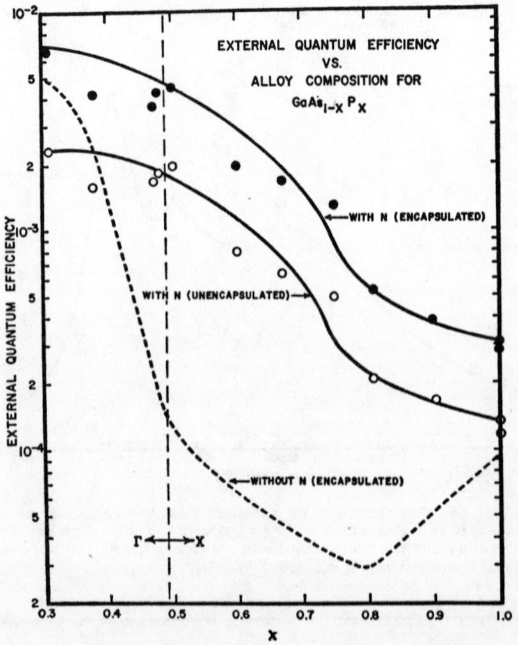

Fig. 12. External quantum efficiency versus alloy composition for $GaAs_{1-x}P_x$ diodes, with and without epoxy encapsulation [5].

will be discussed further in a later section. The following discussion will be concerned with the behavior of diodes with N concentration $\gtrsim 3\times10^{19}/cm^3$.

The quantum efficiency as a function of alloy composition is shown in Fig. 12 [5]. It can be seen that without N doping the quantum efficiency drops very sharply as the direct-indirect transition is approached ($x\sim0.49$ at 300 K) and has a low value throughout the indirect bandgap region where the band-to-band transitions, dominant in the direct bandgap region, are forbidden [3]–[5]. With the addition of N doping, the decrease in efficiency is much less marked as the direct-indirect transition is approached and passed, and the efficiency for alloy composition between $x=0.5$ and $x=0.8$ is more than an order of magnitude higher for the nitrogen-doped diodes than for those without nitrogen doping.

For display applications the brightness and/or luminous efficiency, and not the quantum efficiency, are the important parameters. The brightness takes into account the eye sensitivity as a function of emission wavelength. As shown in Fig. 13 the eye of the so-called standard observer is more than an order of magnitude more sensitive to green emission than to red emission. Therefore, a green-emitting diode will appear much brighter than a red-emitting diode if the quantum efficiencies are equal. The brightness of GaAsP diodes with and without nitrogen doping is shown in Fig. 14. It can be seen that without nitrogen doping the brightness reaches a maximum near $x=0.4$, corresponding to the alloy composition used routinely for commercial red-emitting devices. Throughout the indirect bandgap region the brightness is extremely low. In the case of the nitrogen-doped devices, on the other hand, the brightness reaches a maximum for alloy composition $x\sim0.65$ and remains relatively constant throughout the indirect bandgap region. It is not yet clear whether there is a decrease in the brightness, as indicated by the dashed line in

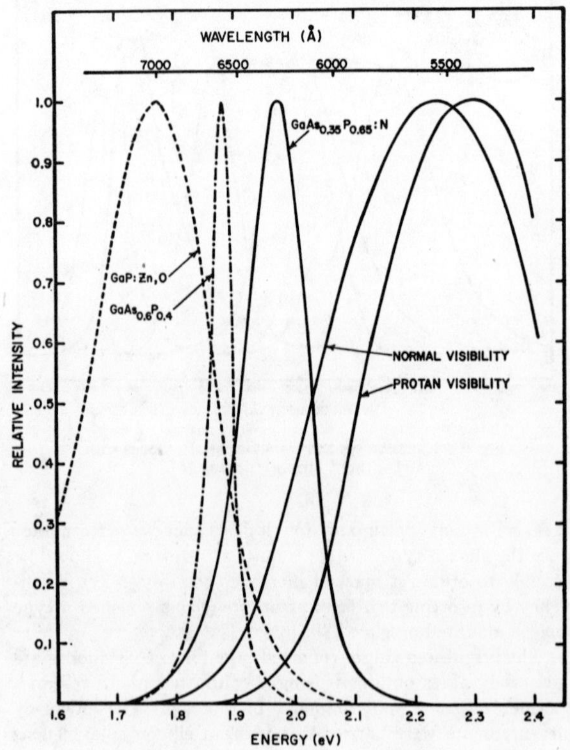

Fig. 13. Eye sensitivity curves for normal and portanopic observers and emission spectra versus wavelength for GaP:Zn, O, $GaAs_{0.6}P_{0.4}$, and $GaAs_{0.35}P_{0.65}$:N red-emitting diodes.

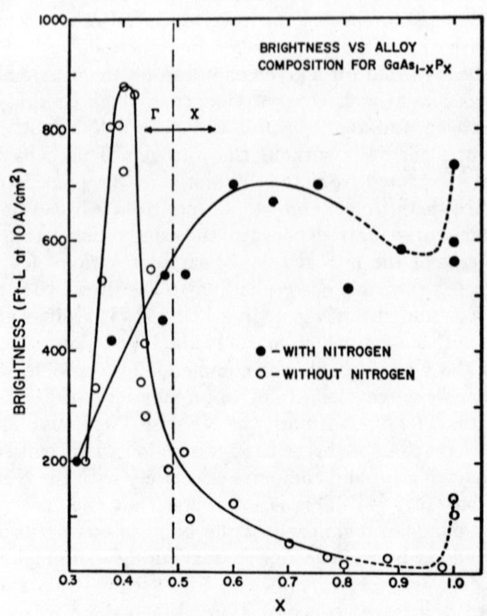

Fig. 14. Brightness versus alloy composition for $GaAs_{1-x}P_x$ with and without nitroben doping [5].

Fig. 14, near $x=1.0$, or if this is simply an artifact due to an insufficient number of data points.

Since the NN pair emission occurs at a longer wavelength than the band edge emission of the nitrogen-free devices, the

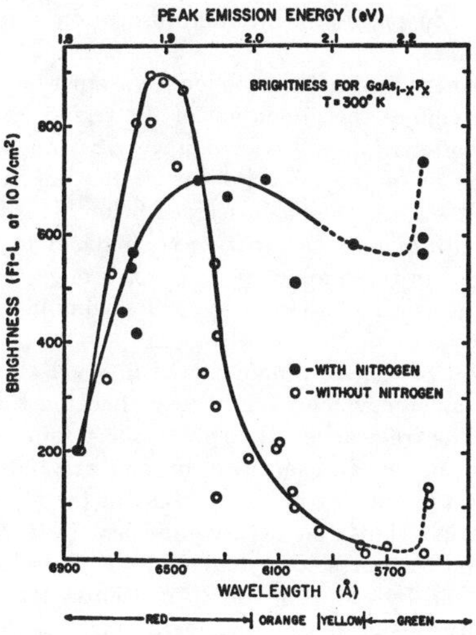

Fig. 15. Brightness versus peak emission wavelength for $GaAs_{1-x}P_x$ with and without nitrogen doping [5].

brightnesses shown in Fig. 14 for a given alloy composition do not correspond to a single color. In order to eliminate this confusing factor, the data of Fig. 14 have been replotted in Fig. 15 which shows brightness as a function of wavelength and peak emission energy instead of alloy composition [5]. Thus the effect of nitrogen doping for the given color of LED can more readily be determined. Clearly, although the nitrogen doping has not yet yielded an improvement in the brightness of red-emitting devices, it has yielded more than an order of magnitude improvement in the yellow and green spectral regions [3], [5]. With nitrogen doping it is now possible to obtain high brightness diodes for all colors from red through green. Further, since the data shown in Figs. 14 and 15 were published [5], improved brightnesses have been observed on nitrogen-doped red-emitting devices. The maximum brightness which has been obtained, for $GaAs_{0.35}P_{0.65}$:N, is 1130 fL at 10 A/cm².

Another factor which should be pointed out is that there are several "critical" $GaAs_{1-x}P_x$ alloy compositions where the direct energy gap Γ is equal in energy to either the indirect energy gap $X(x \sim 0.49)$, the A-line ($x \sim 0.40$), and the NN_3 line ($x \sim 0.36$) [4], [67]. Near these compositions the radiative transition probability is increased which has resulted in the observation, by Holonyak and co-workers [67], of enhanced stimulated emission in optically pumped platelets. Conceivably, diodes and injection lasers with extremely high performance levels could also result.

It should be emphasized that the brightness values quoted here are measured values, not calculated values. The brightness can be calculated using the equation:

$$B = 1150\eta LJ/\lambda \tag{5}$$

where η is the external quantum efficiency, L is the luminous efficiency of the eye in lumens per watt, J is the current density through the junction in amperes per square centimeter, and λ is the emission wavelength in micrometers [9]. However, this equation only applies if the junction area and the emitting area are identical, e.g., that there is no edge emission from the diode. This assumption is generally not that valid since significant edge emission is almost always present, and as a result the equation can be in error by a factor of more than $3x$ when compared to measured values. For example, the sample with x between 0.75 and 0.76 shown in Figs. 12, 14, and 15 has an average calculated brightness of 3000 fL at 10 A/cm² or B/J ratio of 300 fL/A/cm², with a number of individual diodes well in excess of this value. However, the average measured brightness is only 700 fL or a B/J of 70 fL/A/cm², a discrepancy of $4.2x$. This should be remembered when comparing brightness values quoted in the literature since on occasion authors quote only calculated brightness values. For example, calculated B/J values as high as 300 to 320 fL/A/cm² have been quoted for the GaInP alloy system but no comparably high measured values have yet been reported [71].

The brightness which is a surface measurement is a sensible parameter to use to characterize LED's in display applications where the junction area is viewed directly and constitutes the total light-emitting area. However, in an increasing number of applications a LED is used for wide-angle viewing applications, where edge emission is important, or is mounted in a reflective package as described later in Section VI-D, in which case the edge-emitted light is reflected in the forward direction and used to obtain an effectively larger emitting area with a minimum of III–V material. For these applications the total luminous output of a device is a more important consideration than the brightness. The luminous flux, in lumens, is given by the well known expression

$$Lp = 680 \int P(\lambda)\bar{y}(\lambda)d\lambda \tag{6}$$

and the total power output of a light source, in watts, is given by

$$W = \int P(\lambda)d\lambda \tag{7}$$

where $P(\lambda)$ is the power output as a function of wavelength and $\bar{y}(\lambda)$ is the relative photopic luminosity. Therefore, the luminous output per watt output Lp/W can be calculated by the relationship

$$Lp/W = \frac{680 \int P(\lambda)y(\lambda)d\lambda}{\int P(\lambda)d\lambda}. \tag{8}$$

A smooth curve of the lumens-per-watt output for nitrogen-doped GaAsP calculated from a series of diode emission spectra is shown in Fig. 16 [5]. In order to determine the device performance, the lumen-per-watt output data of Fig. 16 must be converted to lumens-per-watt input which is the luminous power efficiency. For most LED devices, with low device impedance, the power efficiency (watts output per watts input) is approximately equal to the quantum efficiency. Thus the luminous power efficiency can be obtained by multiplying the lumens-per-watt data of Fig. 16 by the quantum efficiency of Fig. 12 to obtain the luminous power efficiency data also shown in Fig. 16 [5]. It can be seen that the peak luminous power efficiency of approximately 0.4 lm/W is obtained for alloy compositions in the region near $x=0.65$ with a peak wavelength of approximately 6300 Å. These devices are in the reddish-orange spectral region. The luminous power efficiency of 0.4 lm/W can be compared with commer-

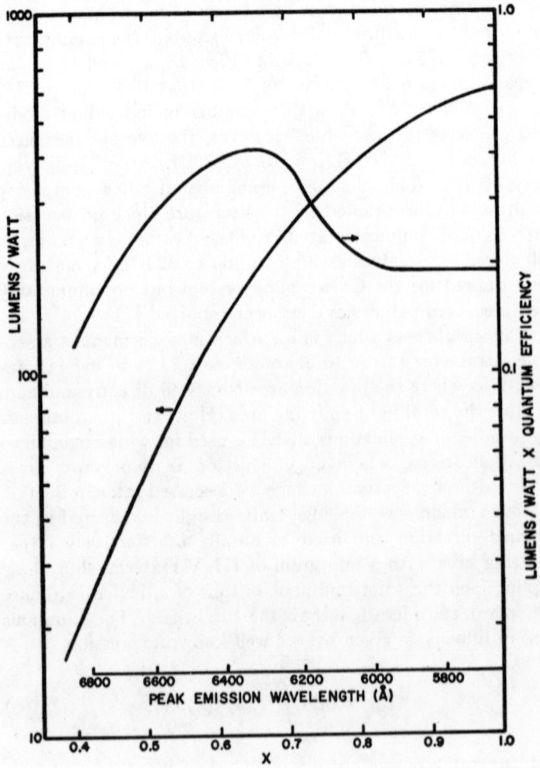

Fig. 16. Luminous output per watt output and luminous power efficiency versus alloy composition for $GaAs_{1-x}P_x$ with nitrogen doping [5].

cial GaAsP devices which typically emit 0.05 lm/W. It should be remembered, however, that there is less disparity in brightness between the commercial GaAsP devices and the nitrogen-doped red emitters discussed here than in luminous power efficiency. The brightness of commercial GaAsP devices is typically 400 fL at 10 A/cm² as compared to 600 to 1000 fL for these devices. The difference is more marked when one compares the luminous power efficiency because in the case of the nitrogen-doped devices the crystal is nearly transparent to the NN pair recombination radiation so that the devices have a vastly increased amount of edge emission. These devices, as is the case with GaP:Zn, O LPE grown junction devices, are well suited to applications calling for wide-angle viewing and mounting in reflective packages.

VI. Device Fabrication Technology and Applications

A. Junction Formation—Zn Profile Considerations

The LED which requires only a single p-n junction is, in principle, an extremely simple device to fabricate. However, the technique used to form the junction is critically important, and for optimum device performance a number of parameters must be considered during the junction formation step or even the best quality starting material will yield devices with low performance levels. This factor has often been overlooked particularly in the case of diffused junctions where the Zn concentration must be carefully controlled in order to avoid damage to the crystal. Factors which must be considered include the minority carrier diffusion length, and the effect of the Zn profile on internal absorption and minority carrier lifetimes.

The minority carrier diffusion length L must be considered since it determines the dimension of the region in which radiative recombination will occur. Since L is proportional to the square root of the minority carrier lifetime it is also a function of the Zn profile, as discussed below. It is important that the depth of the p-n junction is sufficient to avoid the penetration of injected minority carriers through the diffused region to the surface where they will recombine nonradiatively.

The effect of the Zn profile on the internal absorption is another parameter which has a strong effect on the external efficiency. The free carrier absorption due to zinc is roughly proportional to the Zn concentration and can be minimized by reducing the amount of zinc present in the p-type region of the junction. This can be accomplished by reducing the p-type region thickness consistent with diffusion length considerations and by reducing the Zn concentration consistent with minority carrier lifetime considerations.

Finally, the effect of the Zn profile on the minority carrier lifetime is of utmost importance since the internal quantum efficiency is a function of the ratio of the radiative and nonradiative minority carrier lifetimes in the recombination region. For materials, such as $GaAs_{0.6}P_{0.4}$, where free electron-hole recombination dominates, the internal quantum efficiency, η_{int}, can be written:

$$\eta_{int} = \frac{1/\tau_R}{1/\tau} = \frac{1/\tau_R}{1/\tau_R + 1/\tau_{NR}} = (1 + \tau_R/\tau_{NR})^{-1} \quad (9)$$

where τ_R and τ_{NR} are the radiative and nonradiative minority carrier lifetimes, respectively [6], [9]. For materials such as GaP:Zn, O and $GaAs_{1-x}P_x$:N where a trap plays an important role in the recombination, η_{int} is a function of the trap capture rates as well as τ_R and τ_{NR} [72].

In order to maximize the device efficiency one would like to decrease τ_R and increase τ_{NR}. However, τ_{NR} is determined by the quality of the crystal, and the maximum τ_{NR} will be achieved in a crystal with a minimum concentration of nonradiative centers. The nature of the nonradiative centers which limits τ_{NR} in most crystals is generally not well understood although such factors as unintentional background impurities which give rise to deep trap states, stoichiometric defects such as gallium vacancies, and crystal strain and disorder are all important. In any case the important consideration in device fabrication is to avoid any further reduction in τ_{NR} due to the diffusion technique. In general, Zn diffusion results in a surface concentration of Zn atoms of $>4\times10^{19}$/cm³ [6], [73]. If the Zn concentration is not properly controlled an excessive number of dislocations can be generated due to the fact that the Zn atoms are smaller than the host atoms of the crystal lattice [6], [73].

Since τ_{NR} cannot be increased for a given crystal, the ratio τ_R to τ_{NR} can be decreased only by reducing τ_R. The relationship between τ_R and the Zn concentration depends upon the nature of the recombination mechanism. In the case of GaAs and $GaAs_{1-x}P_x$, in the direct bandgap region, radiative recombination occurs as a result of an electron injection into the p-type region followed by recombination with free holes. Thus τ_R can be expressed by the relationship $\tau_R \sim 10^{10}/p$ where p is the hole concentration and 10^{10} is a constant that is in qualitative agreement with the decay times observed in GaAs and other III–V semiconductors [9]. Therefore, by increasing the hole concentration in the recombination region, one can reduce τ_R and increase the internal quantum efficiency. It must be remembered, however, that as the Zn concentration increases there is a tradeoff between reducing τ_R and τ_{NR}. Also, if the Zn concentration is increased to too high of a level, hole injection into the n-type region can occur which is largely nonradiative [74]. The Zn profile which is optimum for one type of material may not be suitable for another. For example, GaAsP generally has a shorter τ_{NR} than high quality GaAs, and thus, in order to minimize the ratio of τ_R to τ_{NR}, it is necessary to form a more abrupt junction with a higher Zn concentration in GaAsP than in GaAs [6].

The considerations discussed here are more important in high-quality crystals where τ_{NR} is long, than in inferior crystals where τ_{NR} is already short. It follows that, as the crystal quality improves, it is important to modify continually the diffusion technique employed in order to insure that optimum device performance results.

B. Diffused versus Grown Junctions

The previous discussion has emphasized the importance of the Zn profile upon the performance of LED's. One can attempt to achieve the optimum profile either by Zn diffusion or grown junction techniques. Both VPE and LPE grown junctions have been investigated, but, to date, VPE grown junctions have not found widespread use in LED applications. LPE junctions, however, are widely used in the fabrication of GAP LED's as well as GaAs IR emitters.

The reputation enjoyed by LPE grown junctions has been enhanced by the fact that LPE grown junctions are relatively easy to form on a research scale, and since the Zn concentration can be controlled by the growth conditions, the technique generally yields good results without the precautions required in the case of diffused junctions. LPE grown junctions are also important since two types of devices, the Si-doped amphoteric GaAs emitters and high-performance GaP:Zn, O emitters can be formed only by LPE growth. Although attempts have been made to duplicate these LPE results using VPE techniques, no success has been reported to date. This is apparently due to the kinetics of formation of the radiative centers, and may be related to the relative concentration of Ga and group V vacancies which would be expected to be quite different in the growth from a gallium melt and from the vapor phase [75].

Although the LPE grown junction is relatively easy to form, and apparently, is the only way to fabricate high-power IR emitters and high-efficiency GaP:Zn, O red emitters, todays commercial market is overwhelmingly dominated by Zn-diffused GaAsP devices. This is largely due to the fact that GaAsP is available in quantity. In addition, however, Zn diffusion has three important advantages with respect to LPE growth. In the first place when the proper diffusion technique is employed, the device performance is both excellent and extremely reproducible, more reproducible than has yet been demonstrated to be the case with LPE devices. The excellent performance which can be obtained using Zn-diffused devices is demonstrated by the results obtained on GaAs Zn-diffused junctions [73]. With high-quality bulk or VPE GaAs and the proper diffusion techniques, diodes have been fabricated with internal quantum efficiencies approaching 100 percent, and external quantum efficiencies as high as any reported using LPE grown junctions with Zn doping. Second, in the case of Zn diffusion, the wafers can be masked so that devices with controlled junction geometry and planar surfaces can be formed. This is particularly important in the fabrication of monolithic numeric and alphanumeric displays. Finally, the diffusion technique is extremely inexpensive when compared to LPE techniques.

C. Wafer Processing

The first step in device fabrication on a VPE $GaAs_{1-x}P_x$ crystal is generally the deposition of a diffusion masking layer, such as Si_3N_4, which is then photolithographically etched to open windows for the junction regions. In some cases, a thin SiO_2 film is deposited in the diffusion window to buffer the diffusion and protect the crystal surface [76].

After masking is completed the wafer is diffused. Occasionally for discrete devices such as those discussed previously in Section V, no diffusion mask is used and the entire surface of the wafer is diffused. The diffusion can be either open-tube [76] or closed-ampoule [2]. A wide variety of diffusion sources have been used such as Zn [33], $ZnAs_2$ [2], Zn–Ga [73], ZnO thin films [77], etc. The proper combination of diffusion source, diffusion time, and temperature must be chosen in order to achieve a suitable junction profile with a minimum of damage to the crystal. Following the diffusion, the wafer is lapped to a thickness of 100 to 200 μm suitable for scribing and dicing, and metallized to form ohmic contacts on both the n- and p-type sides. Typically Au–Ge or Au–Si eutectic alloys are used on the n-type side and Au–Zn or Au–Be on the p-type side. Often, due to the relatively high Zn concentration at the p-type surface the Au–Zn or Au–Be is unnecessary and Al metallization or Al or Au wire bonds can be applied directly to the crystal surface. After metallization the wafers are scribed, diced, and packaged.

D. Applications

The devices which are available commercially include numerics, alphanumerics, and discrete emitters, of a variety of sizes. Many of the devices are available emitting in the red, yellow, or green spectral regions. These include a variety of discrete devices with chip sizes ranging from ~10 to ~50 mils and numerics ranging in height from ~50 to ~600 mils. The numerics may be eigher monolithic or hybrid, and may be packaged in a variety of ways as shown in Fig. 17. The packaging technique is chosen largely on the basis of economic considerations. Generally, only the smallest numerics are monolithic devices, Fig. 17(a), and hybrid techniques, Fig. 17(b) or (c), are employed for larger numerics in order to minimize the use of $GaAs_{1-x}P_x$. The most common numeric, at the present time, is the bar-segment device, Fig. 17(b), in which seven bars of $GaAs_{1-x}P_x$ are used to define the numeric. Rows of junctions are diffused into the bars to form the light-emitting regions. The size and spacing of the junctions is designed to give the appearance of relatively large well-defined bars, with a minimum of current required. In recent months the reflector package, shown in Fig. 17(c), has become increasingly popular, particularly for large numerics, due to its decreased usage of LED material. In this device several small 10–15-mil square chips are attached to a substrate and covered with a metallized plastic reflector which has seven bar-shaped reflector cavities, designed so that a diode is at the center of each cavity. A cross-sectional view of a single cavity is shown in Fig. 17(d). An additional advantage of the reflector package with a diffuser plate is that the light from each chip is distributed over a relatively large area. This can be important

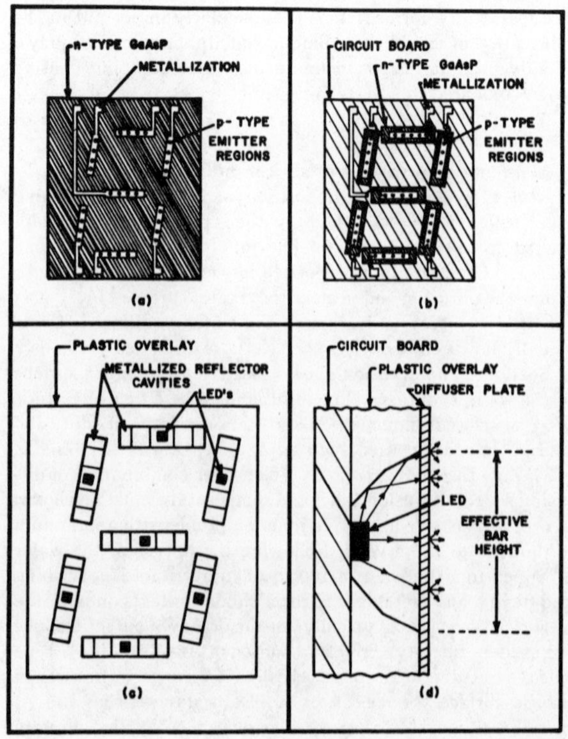

Fig. 17. Schematic drawings of two types of LED numeric packages. (a) A monolithic numeric in which the entire devices is a single chip of $GaAs_{1-x}P_x$. This fabrication technique is generally employed only for small numerics and is often used in conjunction with a plastic magnifying lens. (b) A bar segment device in which a bar of $GaAs_{1-x}P_x$, containing a series of diffused regions, used to define each segment of the numeric. (c) A reflector package. This device employs a small square chip of LED material mounted in a bar-shaped reflector cavity. This type of package which employs a minimum of LED material is economically advantageous for large numerics. (d) Cross section of a reflector cavity.

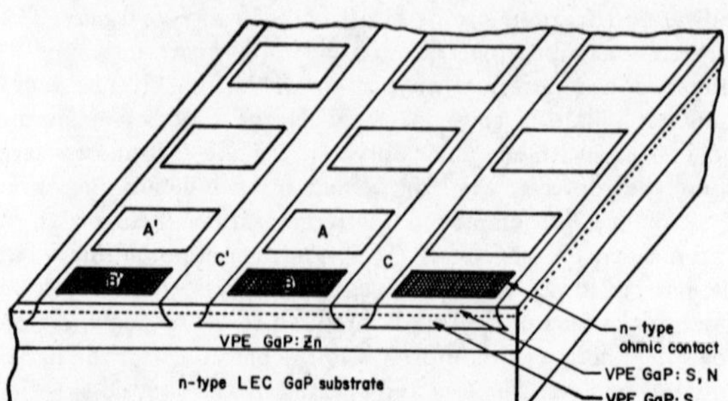

Fig. 18. Schematic drawing of planar electrically isolated GaP array structure [78]. The isolation is accomplished by a deep Zn diffusion through the upper (~50-μm thick) n-type layer of an n-p-n structure. The light is generated at shallow (5–10-μm) junctions diffused into the upper n-type layer.

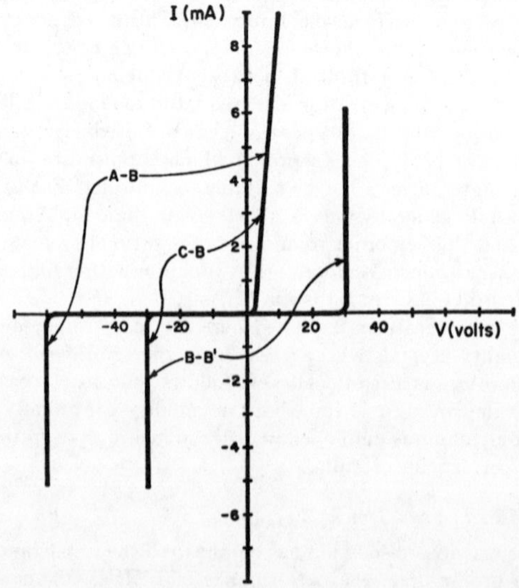

Fig. 19. Typical I–V characteristics for GaP arrays [78]. Curve A–B shows the I–V characteristic of a single-matrix addressable emitter measured between points A and B in Fig. 1. Curve B–C shows the I–V characteristic of the isolation diffusion and buried grown junction. The breakdown voltage for B–C is lower than for A–B because the carrier concentration is higher near the p-type layer. Curve B–B' shows the I–V characteristic between two adjacent n-type channels.

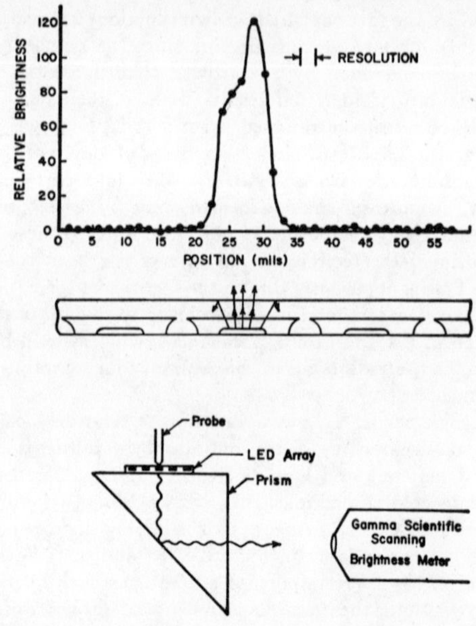

Fig. 20. Optical performance of a typical GaP isolated LED array [78]. The relative brightness is plotted as a function of position on the array surface (viewed through the GaP substrate). The measurement technique is also schematically shown.

because many observers feel that LED's are uncomfortable to look at for long durations due to their high intensity "point-source" emission. As mentioned earlier, this package is somewhat more suitable for GaP:Zn, O and for $GaAs_{1-x}P_x$:N, than for conventional direct bandgap $GaAs_{0.60}P_{0.4}$, since the former have much more edge emission.

Alphanumeric displays are also available, which consist of either 35 discrete chips or a series of bars with a row of diodes in each bar. In either type of device a large number of wire bonds are required resulting in a relatively expensive device. In order to reduce the cost of alphanumeric displays, it is desirable to develop planar monolithic X–Y addressable ar-

TABLE II

PERFORMANCE CHARACTERISTICS OF DIFFERENT TYPES OF LED'S

Device Type	Color	Peak Emission Wavelength Å	lm/W output	Quantum Efficiency		Luminous Power Efficiency	
				Research Result	Commercial Performance	Research Result	Commercial Performance
$GaAs_{0.6}P_{0.4}$	red	6490	75	2.0×10^{-3} [a]	1.0×10^{-3} [b]	0.15	0.075
$GaAs_{0.35}P_{0.65}$:N	red	6320	190	2.2×10^{-3} [c]		0.42	
LPE GaP:Zn, O	red	6925	20	15×10^{-2} [d]	$0.5–2.0\times10^{-2}$ [b]	3.0	0.1–0.4
VPE GaP:N (Zn diffusion)	green	5700	610	0.5×10^{-3} [e]	0.1×10^{-3} [b]	0.3	0.06
LPE GaP:N (grown junction)	green	5700	610	3.0×10^{-3} [f]		1.8	
$GaAs_{0.15}P_{0.85}$:N	yellow	5890	450	0.4×10^{-3} [c]	0.1×10^{-3} [b]	0.18	0.045
VPE GaP:N ($N>10^{20}/cm^3$)	yellow	5900	450	1.0×10^{-3} [e]		0.45	

[a] [2].
[b] Evaluation of commercial devices.
[c] [5].
[d] [8].
[e] [81].
[f] [70a].

Note: The values reported in this table are for dc current densities for $\lesssim 10$ A/cm². Under pulsed high current density operating conditions the nitrogen-doped devices have higher efficiencies. In particular, LPE GaP:N quantum efficiencies as high as 0.6–0.7 percent have been reported, [70a] and [70b].

rays. Devices of this type have recently been reported by two groups of workers [78], [79]. Keune *et al.* [78] have fabricated a monolithic GaP green-emitting $X-Y$ addressable array using a truely planar monolithic process analogous to the technique used in the silicon integrated-circuit technology. A schematic diagram of the array is shown in Fig. 18 with the electrical and optical performance shown in Figs. 19 and 20, respectively. It can be seen in Fig. 19 that the electrical isolation, which is achieved by Zn isolation diffusion through an approximately 50-μm thick n-type layer, is excellent, with leakage currents between adjacent rows of less than 10^{-8}A. The optical performance of this device, which is viewed through the substrate as shown in Fig. 20, is also good with contrast ratios of 200 to 1 being routinely obtained. This large contrast ratio is obtainable since the surfaces of the array chip are planar, which results in internally reflected light being specularly scattered and absorbed or emitted at the edge of the chip rather than through a neighboring emitter region.

The devices are designed to be flip-chip mounted to eliminate wire bonds, and to be edge and end stackable to facilitate the formation of displays with varying formats. Although these devices are presently only in the research stage, all crystal growth and processing steps are designed to be fully compatible with large-scale production techniques. Furthermore, the techniques are suitable for the fabrication of red and yellow displays using $GaAs_{1-x}P_x$:N.

VII. COMPARATIVE PERFORMANCE OF VPE AND LPE DEVICES

In this section the performance of $GaAs_{1-x}P_x$ devices will be compared with the performance of LPE devices for the fabrication of red, yellow, and green emitters. The device performance which has been achieved in the research laboratory and on commercial devices is summarized in Table II. The table includes only devices which are, or should soon be, commercially available. The peak emission wavelength, quantum efficiency (lumens output per watt input) are tabulated for different types of devices of all three colors.

It is immediately apparent that the best research result of 3.0 lm/W for GaP:Zn, O is much better than the performance of any other red-emitting device. However, the typical commercial performance is an order of magnitude lower than the research result, and there is a relatively wide variation in the performance of the commercial devices. This is an indication of the technological difficulties involved in the growth of the LPE devices. This can be contrasted with the case of the $GaAs_{0.6}P_{0.4}$ devices where there is only a $\sim 2x$ difference in performance between the best research devices and the commercially available diodes. The luminous power efficiency of the $GaAs_{0.35}P_{0.65}$:N research diodes is $\sim$0.4 lm/W and, since the growth and fabrication techniques used for the $GaAs_{0.35}P_{0.65}$:N devices is nearly identical to that used for the $GaAs_{0.6}P_{0.4}$ devices, it is reasonable to assume that the typical commercial performance of these diodes should be $\sim$0.2 lm/W which would be competitive with the GaP:Zn, O devices currently available.

In addition to the luminous performance of discrete devices, shown in Table II, other factors should be considered when comparing different types of devices. For example, all of the VPE Zn-diffused devices are compatible with planar processing techniques allowing localized junction geometries, whereas LPE grown junction devices have a p-n junction which covers the entire wafer (localized LPE growth can be achieved, but is not regarded as a viable production technique). While this is not an important consideration for some types of discrete emitters, it does prevent the fabrication of planar monolithic devices. It also makes it much more difficult to employ flip-chip bonding techniques which can, in principle, be employed to eliminate the relatively costly and unreliable wire bonds.

Another performance feature of the VPE $GaAs_{1-x}P_x$:N devices which should be considered when comparing the LPE GaP:Zn, O devices, is the performance under pulsed current conditions. This can be extremely important since the majority of LED display applications use the LED's in a multiplex or pulsed current mode. This mode of operation is well suited to $GaAs_{1-x}P_x$ and $GaAs_{1-x}P_x$:N since both types of devices can be driven to current densities in excess of 500 A/cm² without appreciable loss in quantum efficiency [5]. This is not true of the GaP:Zn, O devices which have a maximum quantum efficiency at low current densities, particularly for high-efficiency diodes [80].

A final performance feature, which can be important particularly for red-emitting devices, is the fact that a certain segment of the population has significantly anomalous color sensitivity. One of the more common class of anomalous observers (approximately 2 percent of the male population) is the protanopic observer who has a decrease in the sensitivity of the eye to the red portion of the spectrum. The eye sensitivity for protanopic observers is shown in Fig. 13. It has been shown that protanopic observers possess approximately 14 percent of the normal photopic sensitivity when viewing red-emitting GaP diodes and about 10 percent when viewing conventional red-emitting $GaAs_{0.6}P_{0.4}$ diodes [81]. However, in the case of the $GaAs_{0.35}P_{0.65}$:N diodes, the protanopic observer would have 28 percent of the normal sensitivity. This is an advantage for the $GaAs_{0.35}P_{0.65}$:N diodes in applications when it is important that the devices have high visibility to the maximum number of observers. Furthermore the red-orange color of the $GaAs_{0.35}P_{0.65}$:N devices, which closely resembles the color of gas-discharge devices, is aesthetically more appealing to many observers than is the deep red of the conventional LED's.

The relative future performance levels and range of applications for $GaAs_{0.6}P_{0.4}$, $GaAs_{0.35}P_{0.65}$:N, and GaP:Zn, O devices is extremely hard to predict at this time. Only a minimal amount of research has been carried out on $GaAs_{0.35}P_{0.65}$:N devices, and with a better understanding of the relevant material growth parameters major improvements in performance may be realized. The performance of commercial $GaAs_{0.6}P_{0.4}$ and the reliability and performance of GaP:Zn, O devices have both improved during the past year and can be expected to continue to improve. The availability of larger diameter GaAs substrates will make $GaAs_{0.6}P_{0.4}$ increasingly attractive from the economic and processing point of view. Because of the inherent differences in cost and performance, various types of red-emitting diodes may coexist in the market place, with each being most suitable for specific applications.

The previous discussion of red emitters concerned three types of devices with different radiative recombination mechanisms, and hence different performance characteristics. In the case of green emitters there is really only one type of device, namely, nitrogen-doped GaP, GaP:N. The only differences in this system are the crystal growth and junction fabrication technique.

Green-emitting LED's can be fabricated by Zn diffusion into n-type LPE or VPE GaP, by LPE growth of the p-n junction, and by VPE growth of the p-n junction, although very little work has been done in the latter area. The best efficiencies reported to date have been for LPE grown junctions, where values as high as ~0.3 percent (for dc currents of ~10 A/cm²) have been reported [70a] with approximately 0.1 percent reportedly attainable on a more reproducible basis. In the case of Zn-diffused devices, efficiencies as high as 0.1 percent have been reported with values in the range of 0.03 to 0.05 occurring more reproducibly [68]. Since quantum efficiency is a strong function of the device-packaging technique, the magnitude of the inherent difference in performance between VPE and LPE green devices has not yet been well established. Brightness is less dependent upon the amount of light which is internally reflected within the chip and finally emitted more or less equally in all directions. For example, both the quantum efficiency and brightness were measured for a grown junction diode fabricated in a high-efficiency mesa package as reported by Logan. The device had a quantum efficiency of 0.06 percent and a brightness on the mesa, which is the brightest area, of approximately 400 fL at a current density of 10 A/cm², or a B/J ratio of 40 fL/A/cm². Thus a grown junction with a quantuum efficiency of 0.1 percent in a similar high-efficiency mesa package should have a brightness of approximately the same as that obtained on the Zn-diffused VPE diodes shown in Fig. 14 although the quantum efficiency is approximately $3x$ higher. This suggests that the grown junction itself may not be markedly superior to the Zn-diffused VPE structure, but that the package enhances the quantum efficiency. A further comparison of VPE and LPE GaP for green-emitting devices has been made in this laboratory by using Zn diffusions and identical packaging techniques for both n-type VPE and LPE material. It has been found that Zn diffusions into n-type VPE and LPE GaP with similar doping levels and nitrogen concentrations, yield devices with equivalent luminous efficiencies [82].

Furthermore, the radiative decay time observed for nitrogen-free VPE GaP diodes is ~400 ns [5] which is equivalent to the best reported values for LPE material [83]. Since the decay time is limited by nonradiative recombination, this suggests that the background concentration of nonradiative centers in VPE and LPE material is approximately the same. This is consistent with the results discussed earlier which demonstrated the excellent electrical properties of VPE GaP [33]. With the addition of nitrogen the radiative decay time has been found to decrease by $\sim 3x$ [5]. Since the radiative decay time is dominated by the nonradiative "shunt" lifetime, this suggests that when the nitrogen is added to the crystal an undesirable defect or impurity which gives rise to nonradiative recombination is also added, resulting in a decrease in the nonradiative lifetime. The fact that the efficiency increases by $\sim 3x$ with the addition of nitrogen as shown in Fig. 12 is consistent with the radiative lifetime associated with the bound N-exciton transition which is $\sim 10x$ shorter than the free-exciton radiative lifetime which gives rise to the emission observed in nitrogen-free GaP [83]. In other words, the nitrogen doping should yield a $\sim 10x$ improvement, as has been observed in LPE GaP [70a], but only yields $\sim 3x$ improvement, presumably due to the simultaneous incorporation of a nonradiative center which decreases the nonradiative shunt lifetime. Thus if the nitrogen incorporation technique can be improved, the efficiency of VPE–GaP:N devices should increase by $\sim 3x$.

Until a comparison of high-quality VPE and LPE devices is carried out using equivalent fabrication and measurement techniques the amount of difference, if any between the two types of devices, will remain unclear. The relative performance of the two types of devices in the future is even less clear since compared to the extensive research effort which has been expended developing red-emitting GaP, a relatively small amount of research effort has been applied to green-emitting devices of either type.

In the case of yellow emitters, there are two VPE techniques that have yielded high-performance room-temperature devices. The first technique is the use of $GaAs_{1-x}P_x$:N as shown in Fig. 15. The other approach is to use VPE GaP:N with N doping greater than $10^{20}/cm^3$. These devices have a yellow color instead of green because as described earlier the emission takes place from the NN pair band instead of from the A-line and its phonon replicas [68], [69]. The performance of yellow devices based upon the two techniques appears to be similar at the present time with brightnesses of 500 to 800 fL at 10 A/cm² and quantum efficiencies of 0.02 to 0.05 percent being attainable [68]. However, in this laboratory it has been found more difficult to grow GaP:N with reproducible electrical and optical properties, than is the case with $GaAs_{1-x}P_x$:N [84]. Another advantage of $GaAs_{1-x}P_x$:N of course, is the fact that by changing the alloy composition the color of the device can be shifted as desired toward either the orange or green spectral region.

There is no good LPE technology for the fabrication of yellow emitters. It has been found that the maximum N concentration which can be incorporated in LPE growth is approximately $3\times10^{19}/cm^3$ [70], in contrast to VPE where greater than $10^{20}/cm^3$ can be readily incorporated [68], [69]. This limitation may be similar to that of observed in the case of oxygen incorporation in GaP [85]. For the case of oxygen it was found that the oxygen concentration in the melt increased with increasing temperature, but the distribution coefficient decreased, with the result that a maximum oxygen concentration of approximately $3\times10^{17}/cm^3$ atoms can be incorporated [85]. Since the VPE growth temperature is well below the approximately 900°C temperature used for LPE, the distribution coefficient may be more favorable for the incorporation of nitrogen. In addition, practically speaking, there is no limitation on the maximum concentration of nitrogen in the gas phase as opposed to the LPE case where the solubility of nitrogen in the Ga melt is relatively low.

Since LPE GaP with an N concentration greater than $3\times10^{19}/cm^3$ cannot be grown, the only LPE technology available for yellow emitters is a combination of the red- and green-device technologies. In devices of this type the n-type region is N doped and the p region is grown with Zn, O doping [86]. This results in a device which emits both red and green light which the eye sees as yellow emission [86]. The relative fraction of green-to-red emission increases with current density since the red emission saturates at high current densities but the green does not. Thus the device changes from red through yellow to green as a function of current [86]. These devices are interesting and their unique property may be of use for certain applications. However, the growth of the devices is complicated and difficult to control reproducibly. Thus it appears unlikely at the present time that this LPE technology will compete with either of the VPE technologies described earlier for the commercial yellow-emitter market.

One other technique which has been used to fabricate yellow-emitting LPE GaP devices should be mentioned. Recently Bhargava *et al.* [87] demonstrated that (Mg, O)-doped GaP emits a weak band of yellow-orange luminescence at 300 K. If the luminous efficiency of these devices can be vastly improved this technique could yield a viable yellow diode.

VIII. Economic Considerations

The LED display technology has the reputation of being a high-performance but also a high-cost technology. However, the cost of LED's has dropped sharply over the past three years and can be expected to continue to decrease. This decrease in cost has been due to improvements in manufacturing techniques, decreasing III–V materials costs, and to the tremendous increase in demand in LED devices in the calculator and other display markets. It is estimated that during 1972 more than 12 million LED numeric displays, and more than 20 million discrete LED's were manufactured [88]. The price erosion has typically occurred at the rate of 20 to 50 percent per year depending upon the point in a given products life cycle, and business conditions [88]. The price "learning" curve

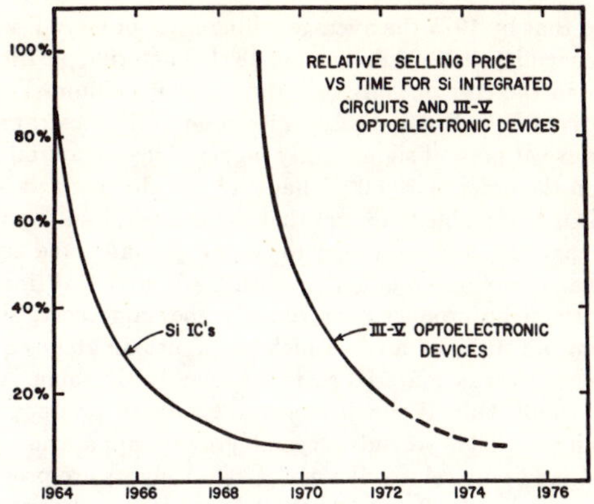

Fig. 21. Relative selling price versus time for Si integrated circuits and III–V optoelectronic devices [88]. The optoelectronic device curve is a combination of the selling prices for a variety of types of LED devices.

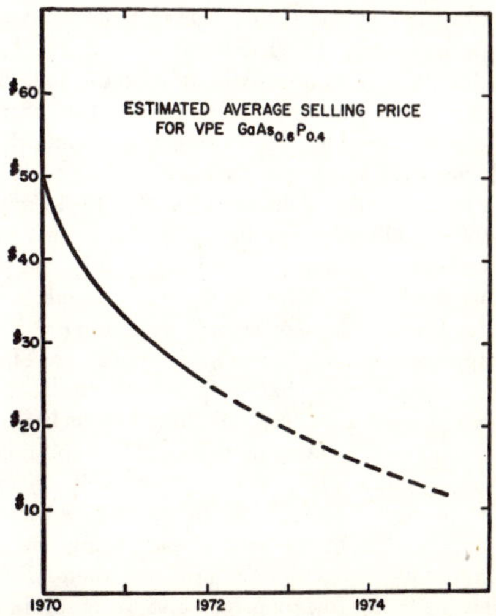

Fig. 22. Estimated average selling price (per square inch) for VPE $GaAs_{0.6}P_{0.4}$ wafers [89]. The price will decrease as substrate areas increases and manufacturing techniques are improved, with the long-range price limited by the cost of gallium.

for III–V optoelectronic devices has very closely followed the curve previously observed for silicon digital integrated-circuit devices as shown in Fig. 21. The decrease in device cost has resulted in continuing pressure on the III–V material suppliers to minimize the cost of the III–V epitaxial wafers. It is the volume demand coupled with the required low material cost that has resulted in the commercial dominance of red-emitting GaAsP LED's.

The relatively well-developed VPE technology has yielded production quantities of material with a high degree of uniformity. It is estimated at the present time that greater than 1.5 million square inches of GaAsP will be used during 1973 and that by 1975 greater than 3 million square inches will be used depending upon the demands of the automobile market [89]. The average selling price of this material during 1972 was on the order of 25.00 dollars/in². The price of GaAsP has decreased steadily over the past few years, as shown in Fig. 22, and can be expected to continue to decrease. It is estimated that by 1975 the average selling price of GaAsP will be in the vicinity of 12.00 dollars/in^2 [89]. The principle limiting factor in the cost of GaAsP is the price of gallium. For the case of VPE GaP and GaAsP grown on GaP substrates the prices are at present significantly higher. These materials now cost on the order of 100.00 dollars/in^2 but the demand is still low. It is reasonable to expect that the price–volume curve for these materials will be similar to GaAsP or GaAs, and assuming that in volume production both GaAs and GaP for substrates can be produced for roughly the same cost, should bottom out at price levels which are approximately equivalent since the raw material costs are roughly the same. At the present bulk GaP is much more expensive to produce since complicated high-pressure crystal growth apparatus is required, and only relatively small (200-g) ingots are produced as compared to 1000-g ingots of GaAs, which can be grown in much less expensive equipment. However, as the bulk gallium phosphide growth technology is improved, larger ingots and higher yields will result and the cost differential will markedly decrease.

Assuming that in the future the cost of epitaxial wafers will be approximately 12.00 dollars/in^2 and the processing cost will double this to approximately 25.00 dollars/in^2, we can make a rough estimate of the possible future cost of LED display devices. Consider, for example, a standard numeric display device with a reflector package as shown in Fig. 17(c) and (d). The chip dimensions are typically 10–15 mils square so that 4400–10 000 chips can be fabricated from 1 in^2 of epitaxial material, which should result in a per chip cost of 0.2–0.6 cents. Thus the 7 LED chips needed for a numeric could be obtained for 1.4–4.2 cents, or even if we assume ~50-percent yield for approximately 3–8 cents per character. This simple calculation shows clearly that, in principle, the cost of LED displays can be extremely low and that there is little validity to the argument which states that since III–V epitaxial wafers are expensive LED's will always be relatively an expensive display technology. The real key to the future cost of LED numeric displays lies in the area of packaging. If labor and package materials costs can be minimized, then LED devices should become quite inexpensive, and it is probable that numeric displays selling for 50 cents per character, or less, depending upon the magnitude of the automotive and other markets, will become available in the future.

During the next few years there will be a cost-effectiveness tradeoff between the VPE growth and Zn-diffusion technology, which dominates the LED market today, and the LPE grown junction technology which is just emerging as a commercial reality. At the present time the performance of the GaP:Zn, O devices is generally adequate, but the GaAsP devices are more reproducible and less expensive. It is likely that the performance of VPE devices, particularly nitrogen-doped GaAsP, will increase markedly over the next few years. It is also likely that an improved LPE growth technique will be developed which will give better reproducibility and which will make more efficient use of raw material, particularly gallium. This will result in a reduced LPE wafer cost, although it is unlikely that the cost will never be as low as for VPE growth. It is not clear at the present time which technology will dominate the future commercial market, but if the performance of VPE devices remains competitive with the LPE devices it will be extremely difficult to unseat the VPE technology which is already firmly entrenched. This is particularly true in view of the compatibility of the VPE growth and Zn-diffusion technology with planar processing techniques similar to those used in the Si integrated-circuit industry. This technology yields high-volume low-cost devices and permits the development of more sophisticated devices such as monolithic numeric and alphanumeric displays.

Packaging, as in the case of LED's, holds the key to the ultimate cost of other display techniques such as liquid-crystal and plasma-panel devices, since also in these devices the cost of the raw material used is minimal. It is likely that in the future a wide variety of display technologies will exist with each technology being most suitable for certain applications on a cost-effectiveness basis. At the present time it is impossible to predict which technique will dominate which section of the commercial market.

IX. Summary

In this paper the VPE materials technology for LED applications has been reviewed with emphasis placed on the commercially important GaAsP and GaP materials. The VPE crystal growth and junction-diffusion device fabrication technologies have been compared to the LPE techniques, both of which can be used to fabricate red-, yellow- and green-emitting LED devices. The principle advantages of the VPE technology are that it is amenable to high-volume production, to the fabrication of devices with reproducible electrical and optical properties, and is compatible with the planar processing techniques already in widespread use in the Si device industry. The latter factor is economically important since it yields minimal device costs, and is technologically important since by controlling the junction geometry with photolithographic techniques one can fabricate more sophisticated monolithic numeric and alphanumeric devices.

In the future VPE materials and devices with lower cost and improved performance can be expected. The lower cost will result from improved manufacturing techniques, larger area substrates, and increasing volumes. For example, it is estimated that more than 3 million square inches of GaAsP will be used by 1975. Improved performance can be expected, particularly in nitrogen-doped $GaAs_{1-x}P_x$, as a better understanding of the effect of various crystal growth parameters is developed.

Acknowledgment

The authors would like to thank J. W. Burd, D. Finn, R. D. Greene, N. Holonyak, Jr. (University of Illinois), and F. V. Williams for helpful discussions, and for supplying some of the data used in this paper. The authors would also like to thank E. Sokoloski for technical assistance.

References

[1] J. W. Burd, "A multi-wafer growth system for the epitaxial deposition of GaAs and $GaAs_{1-x}P_x$," *Trans. Met. Soc. AIME*, vol. 245, pp. 571–576, Mar. 1969.
R. A. Ruehrwein, U.S. Patent 3 218 205, Nov. 10, 1965, assigned to Monsanto Co.

[2] A. H. Herzog, W. O. Groves, and M. G. Craford, "Electroluminescence of diffused $GaAs_{1-x}P_x$ diodes with low donor concentrations," *J. Appl. Phys.*, vol. 40, pp. 1830–1838, Mar. 1969.

[3] W. O. Groves, A. H. Herzog, and M. G. Craford, "The effect of nitrogen doping on $GaAs_{1-x}P_x$ electroluminescent diodes," *Appl. Phys. Lett.*, vol. 19, pp. 184–186, Sept. 1971.

[4] M. G. Craford, R. W. Shaw, W. O. Groves, and A. H. Herzog, "Radiative recombination mechanisms in GaAs P diodes with and without nitrogen doping," *J. Appl. Phys.*, vol. 43, pp. 4075–4083, Oct. 1972.

[5] M. G. Craford, D. L. Keune, W. O. Groves, and A. H. Herzog, "The luminescent properties of nitrogen doped GaAs P light emitting diodes," *J. Electron. Mater.*, vol. 2, pp. 137–158, Jan. 1973.

[6] M. G. Craford, "Properties and electroluminescence of the $GaAs_{1-x}P_x$ ternary system," to be published in *Progress in Solid State Chemistry*, vol. 8. New York: Pergamon, 1973.

[7] R. H. Saul, J. Armstrong, and W. H. Hackett, Jr., "GaP red electroluminescent diodes with an external quantum efficiency of 7%," *Appl. Phys. Lett.*, vol. 15, pp. 229–231, Oct. 1969.

[8] R. Solomon and D. DeFevere, "Efficiency shift in very high efficiency GaP (Zn-O) diodes," *Appl. Phys. Lett.*, vol. 21, pp. 257–260, Sept. 1972.

[9] R. J. Archer, "Materials for light emitting diodes," *J. Electron. Mater.*, vol. 1, pp. 128–154, Jan. 1972.

[10] J. I. Pankove, E. A. Miller, and J. E. Berkeyheiser, "GaN electroluminescent diodes," *RCA Rev.*, vol. 32, pp. 383–392, Sept. 1971.

[11] ——, "GaN blue light-emitting diodes," *J. Luminescence*, vol. 5, pp. 84–86, 1972.

[12] H. P. Maruska, W. C. Rhines, and D. A. Stevenson, "Preparation of Mg-doped GaN diodes exhibiting violet electroluminescence," *Mater. Res. Bull.*, vol. 7, pp. 777–782, 1972.

[13] A. G. Sigai, C. J. Nuese, M. S. Abrahams, and R. E. Enstrom, "Vapor growth of $In_{1-x}Ga_xP$ for p-n junction electroluminescence," presented at 141st. Nat. Meet. of the Electrochemical Soc., Houston, Tex., 1972; also *J. Electrochem. Soc.*, vol. 119, p. 98c, 1972.

[14] J. J. Tietjen, R. E. Enstrom, V. S. Ban, and D. Richman, "Vapor-phase growth of several III-V compound semiconductors," *Solid State Technol.*, pp. 42–49, Oct. 1972.

[15] R. C. Clarke, B. D. Joyce, and W. H. E. Wilgoss, "The preparation of high purity epitaxial InP," *Solid State Commun.*, vol. 8, pp. 1125–1127, Feb. 1970.

[16] D. Richman, "Vapor phase growth and properties of aluminum phosphide," *J. Electrochem. Soc.*, vol. 115, pp. 945–947, Sept. 1968.

[17] J. E. Davey and T. Pankey, "Epitaxial GaAs films deposited by vacuum evaporation," *J. Appl. Phys.*, vol. 39, pp. 1941–1948, Mar. 1968.

[18] ——, "Structural and optical evaluation of vacuum-deposited GaP films," *J. Appl. Phys.*, vol. 40, pp. 212–219, Jan. 1969.

[19] B. Molnar, J. J. Flood, and M. H. Francombe, "Fibred and epitaxial growth in sputtered films of GaAs," *J. Appl. Phys.*, vol. 35, pp. 3554–3559, Dec. 1964.

[20] H. Schafer, *Chemical Transport Reactions.* New York: Academic Press, 1964.

[21] N. Holonyak, D. C. Jillson, and S. F. Bevacqua, "Halogen vapor transport and growth of epitaxial layers of intermetallic compounds and compound mixtures," in *Metallurgy of Semiconduction Materials.* New York: Interscience, 1962, pp. 49–59.

[22] S. M. Ku, "The preparation and properties of vapor-grown GaAs–GaP alloys," *J. Electrochem. Soc.*, vol. 110, pp. 991–995, Sept. 1963.

[23] F. H. Nicoll, "The use of close spacing in chemical-transport systems for growing epitaxial layers of semiconductors," *J. Electrochem. Soc.*, vol. 110, pp. 1165–1167, Nov. 1963.

[24] R. A. Burmeister, Jr., G. P. Pighini, and P. E. Greene, "Large area epitaxial growth of $GaAs_{1-x}P_x$ for display applications," *Trans. Met. Soc. AIME*, vol. 245, pp. 587–592, Mar. 1969.

[25] H. M. Manasevit and W. I. Simpson, "The use of metal-organics in the preparation of semiconductor materials, I. Epitaxial gallium V compounds," *J. Electrochem. Soc.*, vol. 116, pp. 1725–1732, Dec. 1969.

[26] ——, "The use of metal-organics in the preparation of semiconductor materials, II. Epitaxial gallium-V compounds," *Met. Trans.*, vol. 1, p. 623, Mar. 1970.

[27] ——, "The use of metal-organics in the preparation of semiconductor materials, III. Studies of epitaxial III-V aluminum compound formation using trimethylaluminum," *J. Electrochem. Soc.*, vol. 118, pp. 647–650, Apr. 1971.

[28] H. M. Manasevit, F. M. Erdman, and W. I. Simpson, "The use of metal-organics in the preparation of semiconductor materials, IV. The nitrides of aluminum and gallium," *J. Electrochem. Soc.*, vol. 118, pp. 1864–1868, Nov. 1971.

[29] M. Rubenstein and E. Myers, "Epitaxial synthesis of GaAs using a flow system," *J. Electrochem. Soc.*, vol. 113, pp. 365–367, Apr. 1966.

[30] K. L. Lawley, "Vapor growth parameters and impurity profiles on N-type GaAs films grown on N^+-GaAs by the hydrogen-water vapor process," *J. Electrochem. Soc.*, vol. 113, pp. 240–245, Mar. 1966.

[31] C. J. Frosch, "The epitaxial growth of GaP by a Ga_2O vapor transport mechanism," *J. Electrochem. Soc.*, vol. 111, pp. 180–184, Feb. 1964.

[32] J. J. Tietjen and J. A. Amick, "The preparation and properties of vapor-deposited epitaxial $GaAs_{1-x}P_x$ using arsine and phosphine," *J. Electrochem. Soc.*, vol. 113, pp. 724–728, July 1966.

[33] M. G. Craford, W. O. Groves, A. H. Herzog, and D. E. Hill, "Electroluminescence and electrical properties of high-purity vapor-grown GaP," *J. Appl. Phys.*, vol. 42, pp. 2751–2757, June 1971.

[34] A. Mottram, A. R. Peaker, and P. D. Sudlow, "The growth of epitaxial gallium phosphide from the vapor phase by halogen transport," *J. Electrochem. Soc.*, vol. 118, pp. 318–324, Feb. 1971.

[35] C. M. Wolfe, G. E. Stillmann, and J. O. Dimmock, "Ionized impurity density in N-type GaAs," *J. Appl. Phys.*, vol. 41, pp. 504–507, Feb. 1970.

[36] R. W. Shaw, "Influence of substrate temperature on GaAs epitaxial deposition rates," *J. Electrochem. Soc.*, vol. 115, pp. 405–408, Apr. 1968.

[37] W. O. Groves, "The effect of growth orientation on the impurity content of epitaxially grown gallium phosphide," in *Crystal Growth* (*J. Phys. Chem. Solids*, Suppl.) H. Peiser, Ed. (Proc. Int. Conf. Crystal Growth, Boston, Mass., June 20–24, 1966). New York: Pergamon, 1967, pp. 669–672.

[38] J. W. Burd, private communication.

[39] R. A. Burmeister, Jr., and R. W. Regehr, "Epitaxial growth of $GaAs_{1-x}P_x$ on germanium substrates," *Trans. Met. Soc. AIME*, vol. 245, pp. 565–569, Mar. 1969.

[40] O. Igarashi, "Heteroepitaxial growth of GaP on Si substrates by evaporation method," *J. Appl. Phys.*, vol. 41, pp. 3190–3192, June, 1970.

[41] D. A. Grenning and A. H. Herzog, "Dislocations and their relation to irregularities in zinc-diffused GaAsP p-n junctions," *J. Appl. Phys.*, vol. 39, pp. 2783–2790, May, 1968.

[42] M. S. Abrahams, L. R. Weisberg, C. J. Buiocchi, and J. Blanc, "Dislocation morphology in graded heterojunctions: $GaAs_{1-x}P_x$," *J. Mater. Sci.*, vol. 4, pp. 223–235, Mar. 1969.

[43] G. B. Stringfellow and P. E. Greene, "Dislocations in $GaAs_{1-x}P_x$," *J. Appl. Phys.*, vol. 40, pp. 502–507, Feb. 1969.

[44] S. Kishino, M. Ogirima, and K. Kurata, "A cross-hatch pattern in $GaAs_{1-x}P_x$ epitaxially grown on GaAs substrate," *J. Electrochem. Soc.*, vol. 119, pp. 617–622, May 1972.

[45] J. S. Hu, "Misfit dislocations and bowing of wafers in $GaAs_{1-x}P_x$ epitaxy," news paper presented at meet. of the Electrochem. Soc., Miami Beach, Fla., Oct. 8–13, 1972.

[46] A. E. Blakeslee and J. W. Mathews, "Elimination of wafer bending in heteroepitaxial systems," late news item presented at the 2nd Nat. Conf. on Crystal Growth (Amer. Assoc. for Crystal Growth), Princeton, N. J., July 30–Aug. 3, 1972.

[47] G. B. Stringfellow, "The importance of lattice mismatch in the growth of $Ga_xIn_{1-x}P$ epitaxial crystals," *J. Appl. Phys.*, vol. 43, pp. 3455–3460, Aug. 1972.

[48] D. Finn, unpublished data.

[49] E. D. Pierron, private communication.

[50] M. E. Weiner, "Si contamination in open flow quartz systems for the growth of GaAs and GaP," *J. Electrochem. Soc.*, vol. 119, pp. 496–504, Apr. 1972.

[51] W. O. Groves, D. E. Hill, and M. G. Craford, unpublished data.

[52] A. T. Vink, A. J. Bosman, J. A. W. Van der Does de Bye, and R. C. Peters, "Optical properties of excitons bound to neutral Si_{Ga}-donors in GaP and the degeneracy of the Si_{Ga}-donor ground state," *J. Lumin.*, vol. 5, pp. 57–68, Jan. 1972.

[53] M. R. Lorenz and M. H. Pilkuhn, "Luminescence from GaP containing Si," *J. Appl. Phys.*, vol. 38, pp. 61–63, Jan. 1967.

[54] C. E. E. Stewart, "Stoichiometric effects in the growth of doped epitaxial layers of $GaAs_{1-x}P_x$," *J. Crystal Growth*, vol. 8, pp. 259–268, Apr. 1971.

[55] L. J. Vieland, and I. Kudman, "Behavior of selenium in gallium arsenide," *J. Phys. Chem. Solids*, vol. 24, pp. 437–441, Mar. 1963.

[56] R. C. Taylor, J. F. Woods, and M. R. Lorenz, "Electrical and optical properties of vapor-grown GaP," *J. Appl. Phys.*, vol. 39, pp. 5405–5411, Nov. 1968.

[57] H. R. Zwicker *et al.*, "Optical phase shift measurement of residual defects in vapor epitaxial GaAsP," *Solid State Commun.*, pp. 587–590, May 1971.

[58] F. F. Morehead and G. Mandel, "Self-compensation-limited conductivity in binary semiconductors. IV. n-$Zn_xCd_{1-x}Te$," *Phys. Rev.*, vol. 137, pp. A924–A925, Feb. 1965.

[59] G. E. Fenner, "Pressure effect on resistivity of $Ga(As_{1-x}P_x)$," *Phys. Rev.*, vol. 134, pp. A1113–A1118, May 1964.

[60] J. J. Tietjen and L. R. Weisberg, "Electron mobility in $GaAs_{1-x}P_x$ alloys," *Appl. Phys. Lett.*, vol. 7, pp. 261–263, Nov. 1965.

[61] J. Lees, "Mobility variation in GaAs/GaP alloys," *Solid State Commun.*, vol. 6, pp. 11–14, Jan. 1968.

[62] W. O. Groves, unpublished data.

[63] L. M. Foster and J. F. Woods, "Thermodynamic analysis of the III-V alloy semiconductor phase diagrams: I. InSb-GaSb, InAs-GaAs and InP-GaP," *J. Electrochem. Soc.*, vol. 118, pp. 1175–1183, July 1971.

[64] M. Ogirina and K. Kurata, "Effect of donor concentration on several properties of gallium arsenide phosphide," *Jap. J. Appl. Phys.*, vol. 11, pp. 331–337, Mar. 1972.

[65] ——, "Epitaxial growth of $GaAs_{1-x}P_x$," *Jap. J. Appl. Phys.*, vol. 10, pp. 1471–1475, Oct. 1971.

[66] C. M. Wolfe *et al.*, "Effect of the direct-indirect transition on the Hall effect in $Ga(As_{1-x}P_x)$," *J. Appl. Phys.*, vol. 37, p. 434, Jan 1966.

[67] N. Holonyak *et al.*, "Photoexcited resonance-enhanced nitrogen-trap $GaAs_{1-x}P_x$: N laser," *IEEE J. Quantum Electron.*, vol. QE-9, pt. 1, pp. 379–383, Feb. 1973.
R. D. Dupuis *et al.*, "Isoelectronic trap transitions in $GaAs_{1-x}P_x$:N in the region of resonant enhancement ($E_{NN} \sim E_T$, $0.3 < x < 9, 4$)," to be published in *Solid State Commun.*

[68] R. Nicklin, C. D. Mobsby, G. Lidgard, and P. B. Hart, "Efficient yellow luminescence from vapor grown GaP with high nitrogen content," *J. Phys. C*, vol. 4, pp. L344–L347, 1971.
[69] J. W. Burd, "Manufacturing methods for production of single crystal gallium phosphide," Monsanto Co. Interim Rep. IR-515-1(III) of Contract F33615-71-C, Mar. 1972.
[70] a) R. A. Logan, H. G. White, and W. Wiegmann, "Efficient green electroluminescent junctions in GaP," *Solid-State Electron.*, vol. 14, pp. 55–70, Jan. 1971.
b) I. Ladany and H. Kressel, "Efficient green electroluminescent diodes by 'double-bin' liquid-phase epitaxy," *Proc. IEEE* (Lett.), vol. 60, pp. 1101–1102, Sept. 1972.
[71] C. J. Neuse, A. G. Sigai, and J. J. Gannon, "Orange laser emission and bright electroluminescence from $In_{1-x}Ga_xP$ vapor grown p-n junctions," *Appl. Phys. Lett.*, vol. 20, pp. 431–433, June 1972.
[72] R. N. Bhargava, "The physics of radiative centers in GaP," *Phillips Tech. Rev.*, vol. 32, pp. 261–280, 1972.
[73] A. H. Herzog, D. L. Keune, and M. G. Craford, "High-efficiency Zn-diffused GaAs electroluminescent diodes," *J. Appl. Phys.*, vol. 43, pp. 600–608, 1972.
[74] C. J. Nuese, J. J. Tietjen, J. J. Gannon, and H. F. Gossenberger, "Optimization of electroluminescent efficiencies for vapor-grown $GaAs_{1-x}P_x$ diodes," *J. Electrochem. Soc.*, vol. 116, p. 249, 1969.
[75] R. N. Bhargava, S. K. Kurtz, A. T. Vink, and R. C. Peters, "Spectroscopic observation of a vacancy complex in GaP," *Phys. Rev. Lett.*, vol. 27, pp. 183–185, 1971.
[76] K. K. Shih and J. M. Blum, "$GaAs_{1-x}P_x$ electroluminescent diodes made by Zn-diffusion in an open-tube system," *J. Electrochem. Soc.*, vol. 119, pp. 1258–1261, Sept. 1972.
[77] H. Becke, D. Flatley, W. Kern, and D. Stulnitz, "The diffusion of Zn into GaAs to achieve low surface concentrations," *Trans. Met. Soc. AIME*, vol. 230, p. 308, 1964.
[78] D. L. Keune, M. G. Craford, W. O. Groves, and A. D. Johnson, "Monolithic GaP green emitting LED matrix addressable arrays," presented at the IEEE Int. Electron Devices Meet., Washington, D. C., Dec. 1972.
[79] D. Digvet and J. Lebailly, "$GaAs_{1-x}P_x$ and $Ga_{1-x}Al_xAs$ monolithic crossbar displays," presented at the IEEE Int. Electron Devices Meet., Washington, D. C., Dec. 1972.
[80] W. H. Hackett, Jr., W. Rosenzweig, and J. S. Jayson, "Saturation of Zn-O complexes in GaP diodes," *Proc. IEEE* (Lett.), vol. 57, pp. 2072–2073, Nov. 1969.
[81] M. R. Allyn, R. W. Dixon, and R. Z. Bachrach, "Visibility of red and green electroluminescent diodes for color-anomalous observers," *Appl. Opt.*, vol. 11, pp. 2450–2454, Nov. 1972.
[82] A. H. Herzog, unpublished data.
[83] R. Z. Bachrach and O. G. Lorimor, "Measurement of the exterior room-temperature minority carrier lifetime in GaP," *J. Appl. Phys.*, vol. 43, pp. 500–507, 1972.
[84] W. O. Groves and D. Finn, unpublished data.
[85] R. H. Saul, "Effect of growth temperature on oxygen incorporation in GaP," *J. Electron. Mater.*, vol. 1, pp. 16–23, Jan. 1972.
[86] W. Rosenzweig, R. A. Logan, and W. Wiegmann, "Variable hue GaP diodes," *Solid-State Electron.*, vol. 14, pp. 655–660, 1971.
[87] R. N. Bhargava *et al.*, "Mg-O complexes in GaP-A yellow diode," *Appl. Phys. Lett.*, vol. 20, pp. 227–229, 1972.
[88] R. D. Greene, private communication.
[89] F. V. Williams, private communication.

Nematic Liquid Crystal Materials for Displays

LINDA T. CREAGH

Abstract—Nematic liquid crystals are being used in a variety of display devices that capitalize on unique characteristics of these organic fluids. This paper discusses the purification and preparation of nematic liquid crystals, their operating temperature ranges, and the extent to which their properties meet the requirements of current dynamic-scattering and field-effect displays. While device performance can be further improved by tailoring the material, product acceptance will be principally determined by other factors such as reliability, packaging, and price reduction.

LIQUID CRYSTALS have been investigated by chemists and physicists for many years. They remained, however, essentially laboratory curiosities until the realization of their potential application in display devices. The fundamental concepts of the interaction of certain classes of liquid crystals with electric, magnetic, and ultrasonic fields were recognized by Bjornstahl [1] and Chistyakov [2], but the actual implementation and reduction to practice was not accomplished until Fergason's group [3] at Westinghouse and Heilmeier and co-workers [4] at RCA began work. It was at this time that the unique characteristics of nematic liquid crystals really began to be exploited.

The purpose of this paper is to explore the utilization of nematic liquid crystals in displays. The primary emphasis will be on the applicablity of characteristics of the organic material, the extent to which liquid crystals presently meet display demands, and the prospects of improving liquid crystals so as to better satisfy a useful or practical customer requirement. The molecular structure or chemical basis of liquid crystals will not be discussed because of the availability of recent extensive reviews [5].

Readers unfamiliar with the construction and operation of liquid-crystal displays might refer to Heilmeier's early publications or to "Performance Characteristics of Nematic Liquid Crystal Display Devices" [6] for review since the subsequent discussion is limited to the role of the liquid crystal itself in devices.

Since Heilmeier's pioneering work [4], several electro-optical phenomena have been used in devices. The following list does not include all the effects that have been observed for liquid crystals, but mentions those that appear to be most applicable to device development: 1) dynamic scattering: a) ac drive, b) dc drive; 2) field effect: a) twisted nematic, b) tunable birefringence, c) cholesteric–nematic phase change; 3) thermal addressing: a) cholesteric, b) smectic.

For field-effect modes, there is considerable variation in terminology. The nomenclature used here is employed for clarity and is not intended to sanction any particular label. Twisted nematic devices are so named because of the initial ordering of the liquid crystal [7]; on the other hand, tunable birefringence [8] refers to the change in refractive index on application of electric field with homeotropically aligned nematic fluids having negative dielectric anisotropy. DAP (distortion of aligned phases) [9] and forced alignment techniques [10] are other terms that have been attached to the latter effect. At the 4th International Liquid Crystal Conference (Kent, Ohio, Aug. 1972), there was mention by several session chairmen of the applicability of the "Freederickz effect" [11] for field-assisted orientation changes. Unfortunately, this term does not discriminate well between effects which differ in properties and applications; it is too general.

Preparation and Purification

Liquid crystals in common use are relatively simple organic compounds or mixtures of such compounds. A diverse group of long rod-like molecules are known to be nematic mesomorphs, but only a few types of these compounds have proved feasible in devices. These compounds are prepared by standard techniques of organic chemistry. Typical routes for the most important classes of compounds are indicated in Fig. 1.[1]

The products of these reactions can be purified by classical organic methods. Multiple recrystallizations are valuable in removing by-products, but recrystallization is not always successful, as discussed by Steinstrasser [12] in the attempt to synthesize *p*-*n*-heptylphenyl-*p*′-acetoxybenzoate (Reaction 1, Fig. 2). Five recrystallizations gave a monotropic liquid crystal with a melting point of 53°C and a clearing point of 51°C. Final purification was accomplished by preparative layer chromatography, and the material was monotropic with a melting point of 53°C and a clearing point of 47.5°C. The by-product is nematic from 106–215°C.

Chromatographic isolation has provided a route to isolation of milligram-size samples of isomeric liquid crystals [13] (Reaction 2, Fig. 2). Oxidation of azobenzenes yields a mixture of two isomers that are difficult to separate other than chromatographically. The product being a mixture has a nematic range broader than either individual component, and as a result the mixture may be more applicable to device use than either single compound.

Many liquid crystals can be gas-chromatographed analytically, and this can in principle be extended to preparative scale. Princeton Organics sells several types of liquid crystals which have been purified by "vacuum sublimation," having found that if a liquid crystal can be analyzed by gas chromatograph (GC), it can be vacuum-distilled [14]. Vacuum distillation can in the case of 4′-methoxybenzylidene-4-butylaniline (MBBA) give a product that will show only one peak when analyzed by GC and have a transition temperature of 47°C [15], [16].

Zone refining has been applied in certain instances [17]. Haberfeld and co-workers [18] have found zone refining to increase purity (as determined by differential scanning calorimetry) compared to solvent recrystallization; however, Carr found zone refining unsuccessful when applied to ethyl *p*-[(*p*-methoxybenzylidene)animo]cinnimate [17].

Manuscript received January 3, 1973; revised January 24, 1973.

The author is with Texas Instruments Inc., P.O. Box 5936, Dallas, Tex. 75222.

[1] References are not intended to be the original synthetic route, but do give fundamental information on synthesis and purification.

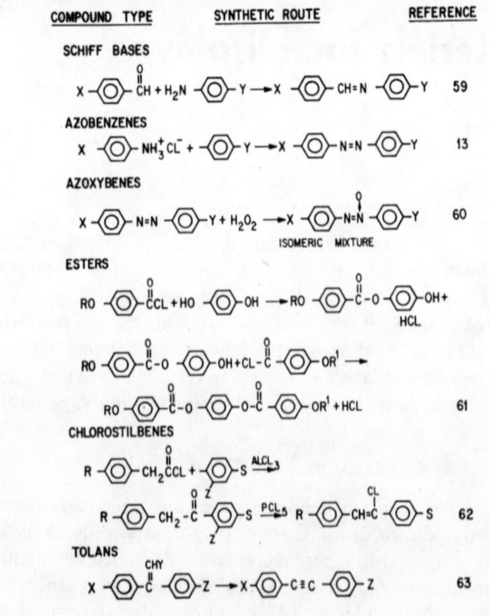

Fig. 1. Typical methods for synthesis of nematic liquid crystals.

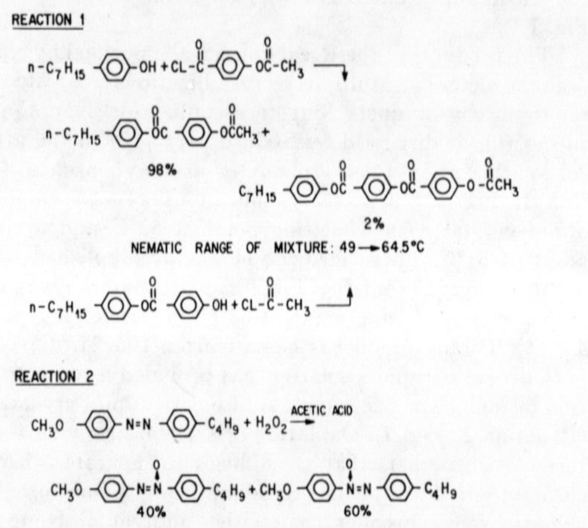

Fig. 2. Synthesis pathways yielding mixtures of liquid crystals. Reaction 1: Two different methods produce a mixture of esters in which the minor constituent has a high nematic temperature range and the major component is monotropic [12.[Reaction 2: Oxidation of azobenzenes yields roughly equal amounts of the two possible isomers [13].

Some of the important materials characteristics for displays are determined not by the degree of organic contamination but by traces of ionic materials and water. As a result, the purity of liquid crystals should be evaluated on both points.

The removal of ionic species is of especial significance for field-effect devices and for fundamental studies in dynamic scattering, electrochemical processes, and reliability. The general concepts for improving resistivity have been based on experiments with organic liquids such as nitrobenzene [19]. It has been shown [20] that conventional methods such as vacuum distillation and electrolysis do not reduce ionic contamination as efficiently as ion exchange techniques. Electrodialysis using ion exchange membranes has been used [19], [21] to produce nitrobenzene with a resistivity of 1.7×10^{13} Ω·cm, compared to 10^8 Ω·cm initially. These techniques have been extended to MBBA [22] resulting in changes in resistivity of 3×10^9 Ω·cm to 2×10^{12} Ω·cm. By way of comparison, Kashnow and Cole [23] found it possible to obtain MBBA with a resistivity of 5×10^{11} Ω·cm by multiple recrystallizations from absolute ethanol, vacuum distillation, and then storage over a molecular sieve. Material thus prepared has a dynamic scattering cutoff frequency of 5 Hz.

The level of purity required of the liquid crystal is dictated by the application. Several operating parameters are limited by the resistivity, whereas the reproducibility of transition temperatures is determined by the level of organic impurities.

TABLE I

TEMPERATURE REQUIREMENTS FOR LIQUID-CRYSTAL DEVICES

Application	Acceptable Temperature Range (°C) Storage	Operating
Commercial calculators, clocks, instrument panels, multimeters, industrial controls	−45° to 70°	0° to 45°
Watches	0° to 55°	8° to 48° wrist temperature of 28°
Government products	−55° to 90°	−55° to 70°
Automotive products	−55° to 100°	−55° to 70°

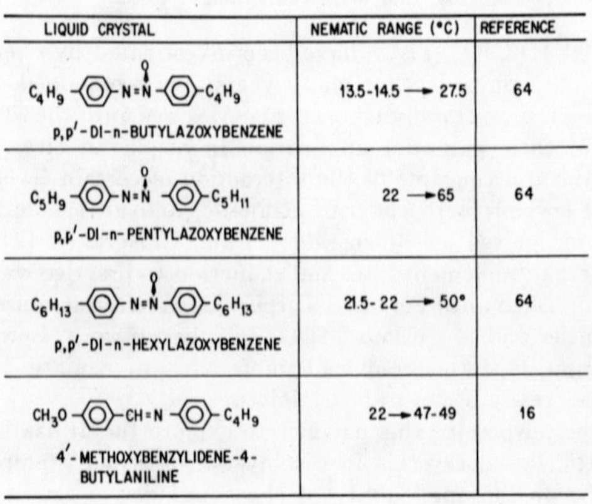

Fig. 3. Examples of individual liquid crystals that are reported to be nematic at room temperature.

TEMPERATURE RANGE

For any of the known electrooptical effects to be truly useful, the operating temperature range of the display must be of sufficient breadth. Table I gives an indication of the needs of device customers.

The temperature ranges over which display devices will function are limited by the mesomorphic range of the liquid-crystal materials. Even though external heating may be possible, the obviously advantageous materials for devices are those that are mesomorphic under ambient conditions. In 1969, Kelker and Scheurle [16] made the first report of a compound, MBBA, with a nematic range encompassing room temperature. Today several room temperature nematics are known (see Fig. 3), but the usable temperature range of individual compounds is limited.

The principal means of obtaining an adequate working

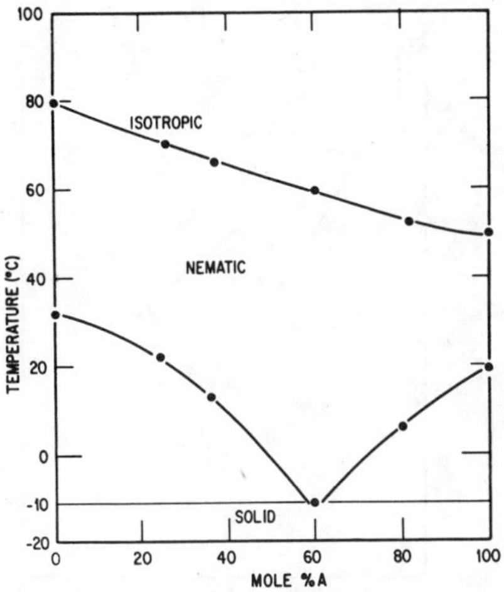

Fig. 4. Binary phase diagram for MBBA (component A) and EBBA (component B).

temperature range is through the use of mixed systems with one or all constituents nematic mesomorphs. Fig. 4 gives the binary phase diagram for MBBA:EBBA (4′-ethoxybenzylidene-4-*n*-butylaniline). This illustration is typical for nematic liquid crystals in that the nematic-to-isotropic transition temperature changes uniformly and there is a rather abrupt minimum at the eutectic point. It should be noted that liquid crystals readily form supercooled liquids and some care is necessary in determining the crystalline-to-nematic temperature.

The ease of formation of a metastable liquid is affected to a marked extent by material purity. A binary of MBBA:EBBA (60 mole percent: 40 mole percent) with total organic impurities of 1 percent readily supercools to −20°C in 12-μm sandwich cells; after the mixture finally crystallizes, the solid-to-nematic transition temperature is −14°C. The same experiment performed with MBBA:EBBA with total organic impurities of less than 0.1 percent shows that supercooling is difficult to observe and that the crystalline-to-nematic transition temperature is −10°C.

Ternary [24] and quaternary [12] mixtures of nematic liquid crystals can also be used to obtain wide ranges of operation and storage. These complex solutions tend to form glasses rather than to crystallize on cooling [25]. Such materials can exist for prolonged periods of time below the freezing point without crystallizing. Petrie [26] has suggested that the temperature cycling of binaries and higher order mixtures that are not eutectic may result in segregation through preferential crystallization of the excess components. Although such effects could be reversible and the excess material would redissolve, the rate of return to homogeneous composition may be slow and crystallite patterns may distort the original alignment.

Nematic mixtures suitable for use in display fabrication are available from many sources. Table II lists some of the materials along with the manufacturer's specified temperature range. None of these liquid crystals fill all the needs of device customers.

Information on temperature range and resistivity in Table II is taken from suppliers' data sheets. These numbers should be rechecked if they are critical to device performance, as there is variation from sample to sample.

TABLE II

SOME COMMERCIALLY AVAILABLE NEMATIC LIQUID CRYSTALS

Liquid Crystal	Manufacturer or Supplier	Nematic Range (°C)
Tolan binary mixture	Anvar, France	0–59
Nematic Mixt. #11643	Eastman Kodak	15–97
MBBA #11246	Eastman Kodak	19–45[a]
#36320	Riedel-DeHaen AG, Germany	21–44–45
	VARI-LITE	10–48
	Princeton Organics	22–48[b]
Chlorostilbenes #16,407-0 and 16,408-9	Aldrich Chemical Co.	8–59[c]
Licristal Phase 4(IV)	E. Merck	16–76
Phase 5(V)		−5–75
Phase 5A(VA)		−5–75
Phase 6		−10–80
LC250	Princeton Organics	−5–65[d]
LC260		15–85
LC350		15–75[e]
Nematic liquid crystal	Liquid Crystal Industries	18–80
Nematic liquid crystal		0–70[f]
VL-(-20)55-N	VARI-LITE	−20–55[g]
VL-060-N	VARI-LITE	0–60[h]

[a] 6×10^9 Ω·cm resistivity at 25°C.
[b] 5×10^{10} Ω·cm resistivity at 25°C.
[c] 40–60 mole percent mixture to obtain temperature range [31]. Very light sensitive.
[d] 5×10^9 Ω·cm resistivity at 25°C.
[e] 10^{11} Ω·cm resistivity at 250°C, $\epsilon_A>0$.
[f] $\sim10^{10}$ Ω·cm resistivity.
[g] 4×10^9 Ω·cm resistivity.
[h] 1.2×10^{11} Ω·cm resistivity.

What is the expectation that nematic liquid crystals will be available that will meet the most stringent demands of any customer? The bottom temperature (freezing point) is more difficult to achieve than the top (nematic-to-isotropic). The nematic-to-isotropic point for some classes of liquid crystals is quite high, as shown in Fig. 5, so one can envision use of proper mixing to obtain nematic-to-isotropic transitions in excess of 100°C. Fig. 6 gives the phase diagram for an ester binary in which the nematic-to-isotropic transition temperature is about 290°C. That these compounds can be quite thermally stable is shown by their use as stationary phases in gas chromatography [27].

As far as storage in the upper temperature range is concerned, it appears that exceeding the nematic-to-isotropic transition temperature is a packaging problem rather than a liquid crystal problem. It is possible to contain pure MBBA and pure EBBA in glass under vacuum so that there is no change in chemical characteristics after 13 000 h at 85°C (see Fig. 7). Other types of liquid crystal, being more chemically stable than Schiff bases, should behave at least as well. If package integrity is assumed, the cycling of the liquid-crystal device through the nematic-to-isotropic point generally does not destroy aligning forces so that the original alignment is preserved on cooling.

The use of bulk surfactants complicates the chemical picture for high-temperature storage. Preferential adsorption of the surfactant to the substrate may be altered by increasing temperature. Chemical reactions may occur between the liquid crystal and the additive, or the additive may be thermally unstable itself. The progress of these complex chemical reactions can be followed by the lowering of the nematic-to-isotropic transition temperature of MBBA, as shown in Fig. 8, where samples of MBBA were deliberately doped with starting materials [15]. Experiments at Texas Instruments have shown that while MBBA alone is thermally stable, MBBA saturated with hexadecyltrimethylammonium bromide grad-

LIQUID CRYSTAL	NEMATIC RANGE (°C)	REFERENCE
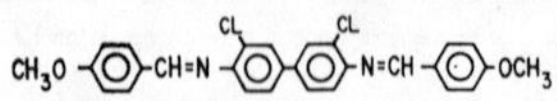 4,4'-BIS(p-METHOXYBENZYLIDENEAMINO)-3,3'-DICHLOROBIPHENYL	154→334	65
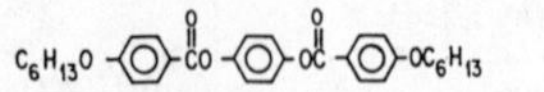 1,4-PHENYLENE BIS (4'-HEXYLOXYBENZOATE)	124.4→212.9	66
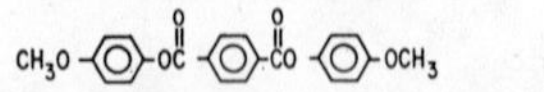 DI-p-METHOXYPHENYL TEREPHTHALATE	205→277	67
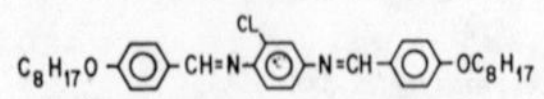 BIS(4'-OCTYLOXYBENZAL)-2-CHLORO-1,4-PHENYLENEDIAMINE	59→179	68

Fig. 5. Examples of nematic liquid crystals with wide temperature ranges. Such materials can be used to achieve mixtures that are nematic from below 0°C to above 100°C.

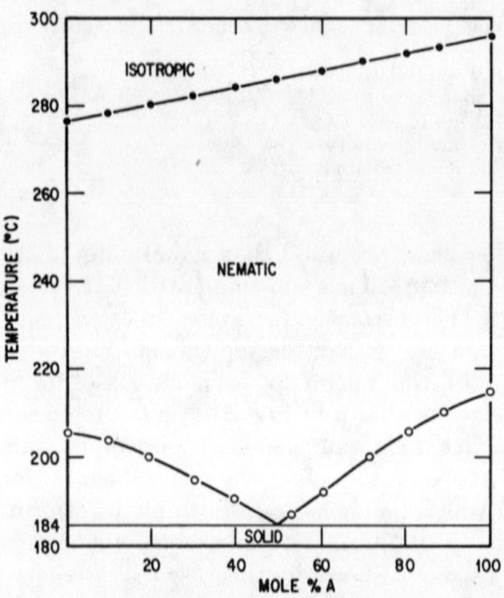

Fig. 6. Binary phase diagram for an ester system. Component A: *p*-phenylenedi-*p*-anisic ester; component B: di-*p*-methoxyphenyl terephthalate. (From Dewar and Goldberg [18].)

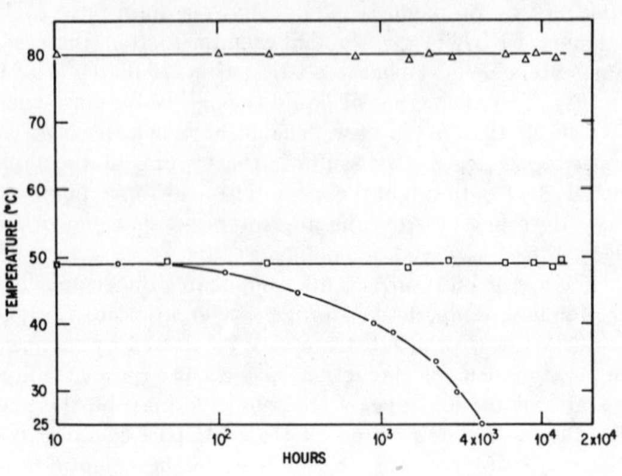

Fig. 7. Nematic-to-isotropic transition temperature as a function of time for EBBA (△), MBBA (□), and MBBA saturated with hexadecyltrimethylammonium bromide (○). The samples were sealed *in vacuo* in clean glass ampoules and stored at 85°C with minimum light exposure. Transition temperatures were monitored on a heated microscope state stage with accuracy ±1°C.

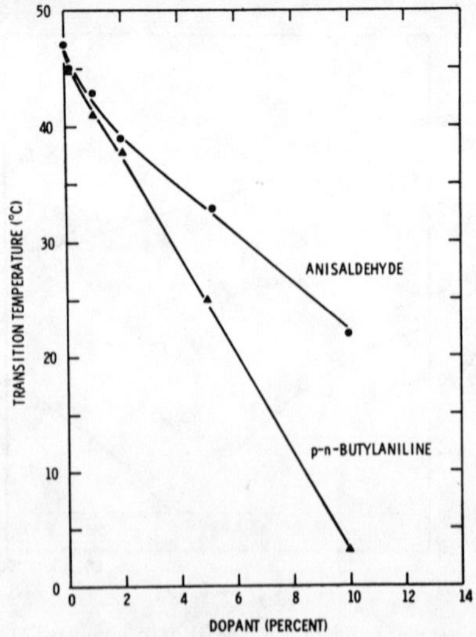

Fig. 8. Change in nematic-to-isotropic transition temperature of MBBA with added *p*-anisaldehyde (●) and *p*-*n*-butylaniline (▲). (From Creagh *et al.* [15].)

ually degrades at 85°C, as shown on Fig. 7. In general, this upper transition temperature is quite sensitive to impurities. In the case of the quaternary bromide, discoloration also occurs.

The concept of liquid-crystal displays functioning adequately at low temperatures requires the evaluation of several factors. For field effects [28] and dynamic-scattering devices [29], the response times of liquid crystals have been shown to approximately fit equations of a form

$$\tau_{\text{rise}} = \frac{C_1 \eta d^2}{\epsilon V^2} \tag{1}$$

$$\tau_{\text{decay}} = \frac{C_2 \eta d^2}{K} \tag{2}$$

where C_1 and C_2 are constants characteristic of the liquid crystal; η is the viscosity; ϵ is the effective dielectric constant; K is a Franck elastic constant; d is cell thickness: and V is the applied voltage. The time of response to electric field or voltage is proportional to the viscosity of the liquid. Berchet *et al.* [30] have determined viscosity as a function of temperature for several liquid crystals and found a range of 25–30 cp MBBA (22–45°C) and 1–8 cp for liquid crystals with higher nematic temperature ranges (~60–175°C). Extrapolation of such data back to very low temperatures explains the observed response times for actual displays. At −15°C the turnon time is of the order of seconds, even at 100 V dc, and the decay is at least as long. To meet time requirements it appears that heating to some temperature like 25°C will be necessary (specific temperature depends on application).

The use of mixtures to meet storage requirements offers more promise, but military specification of −55°C will be difficult to achieve. Even if one considers materials with low molecular weight, relatively few organic materials are liquid at this temperature. The compounds that are fluid have weak intermolecular forces in contrast to the strong attractions found for liquid crystals [31]. Consequently, it is difficult to visualize a molecule possessing the structural features necessary to be liquid at low temperatures and be a liquid crystal

TABLE III

THERMAL EXPANSION CHARACTERISTICS

Material	Coefficient of Thermal Expansion (°C^{-1})	Reference
MBBA	0.807×10^{-3} (nematic)	
PAA	0.72×10^{-3} (nematic)	[32]
	0.75×10^{-3} (isotropic)	
Licristal IV	0.67×10^{-3}	[33]
Licristal V	0.64×10^{-3}	
Licristal VA	0.64×10^{-3}	
Typical orange fluids	$0.2\text{–}1.6\times10^{-3}$	[34, p. 2067]
Glasses: soft, plate, hard	$8.5\text{–}9.7\times10^{-6}$ (linear)	[34, p. 2065]

above room temperature. Unless glass formation in the liquid crystal and subsequent thawing are acceptable, some minimum heat will need to be supplied to the inactive display in order to meet storage requirements.

The significance of freezing during storage must be evaluated. The organic material will not itself be damaged by cold; in fact, any degradation processes due to chemical change will be retarded by low temperatures. If the package can withstand temperature cycling without bubble or void formation in the liquid or breaking of the seal, storage below the freezing point would be acceptable and one would only have to tailor the heating system so that the liquid crystal would be thawed and sufficiently warm to respond properly when desired. However, the initial alignment of the liquid crystal may be affected by freezing, especially if crystals form rather than a glasslike structure. In some cases, the pattern of the crystal is retained even after cycling through the nematic-to-isotropic point. This reduces the contrast in a dynamic-scattering device and can destroy the high degree of order essential to a field-effect device. Where alignment retention is important, it may be necessary to thermostat the display environment at temperatures sufficient to prevent crystallization.

Packaging aspects of the temperature problem should not be dismissed lightly. The package must be designed with consideration given to the large coefficient of expansion characteristic of liquid crystals. Table III gives volume coefficients of thermal expansion for several liquid crystals and for some typical glasses. Also, contraction and expansion tend to cause voids or bubble formation in the liquid-crystal area which may result in cosmetic failure. It is especially significant if polarizers are needed in the display.

Photochemical degradation must be considered for the liquid crystal-additive system if any component is light sensitive, and precautions may be necessary to preclude photochemical reaction. For example, Hampel reported [35] that Licristal IV, V, and VA show a gradual increase in response time under illumination due to photochemically induced material changes. The addition of 10 percent of a new nematic ester was found to retard this process to an acceptable extent. Chlorostilbenes have been found to be sensitive to UV radiation passed by soft glass. Haller [38] has developed a clear coating for the outer glass surfaces to prevent these deleterious reactions.

Assuming that the operating and storage temperature characteristics are satisfactory, then further material specification is required to meet the demands of the particular electrooptical effect to be used in the display. No one material can be designed to fulfill all the demands of all applications. In the following discussion, emphasis will be placed on what

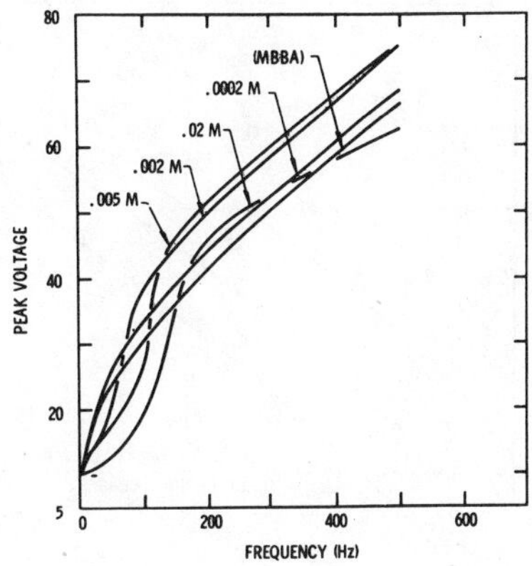

Fig. 9. Frequency dependency of threshold voltage for MBBA and MBBA doped with decyltrimethylammonium bromide to the indicated molar concentrations. (From Creagh *et al.* [15].)

can be done to the liquid crystal itself to determine display characteristics.

DYNAMIC SCATTERING—AC

Light scattering in liquid crystals is due to microscopic variations in the refractive index accompanying local disruptions of the molecular alignment caused by turbulence in the liquid. The turbulence can be induced by the application of electric fields. When the electrohydrodynamic instabilities arise under ac drive, the phenomenon is explained by the Carr–Helfrich theory [37], and when turbulence results from dc bias, the instabilities generally fit the analysis suggested by Felici [30].

For effective dynamic light scattering, the nematic fluid is required to have negative dielectric anisotropy [39]. In dynamic-scattering displays using ac drive, it is necessary to prepare material with the proper resistivity because the voltage necessary to achieve adequate contrast is related to the drive frequency [23] (see Fig. 9). Changes in conductivity are commonly accomplished by starting with liquid crystals of high resistivity and dissolving in them ionic compounds such as quaternary ammonium salts. Although many of these salts have limited solubility in pure dry liquid crystal, orders of magnitude change in resistivity can readily and reproducibly be accomplished [40]. For multiplexed drives using two ac frequencies as described by Stein and Kashnow [41] at General Electric, material with a high cutoff frequency is necessary. Relatively highly conductive material can also be used for low-frequency dc drive.

Applications requiring low power drain, such as watches where battery life of one year is specified by the customer, necessitate the reduction of ac current to a minimum. A desirable current level appears to be about 2 $\mu A/cm^2$ of active display area. Thus far, manufacturers have found it difficult to produce displays with such low current and sustain adequate contrast at 15–20 V_{rms}.

One major problem of low-current operation is the decrease in ac current with decreasing temperature [42], as shown in Fig. 10. It should be noted that light scattering effectively ceases when the current falls below 0.5 $\mu A/cm^2$ (15 V_{rms}, 32 Hz). Although the data presented are for Schiff bases doped with quaternary ammonium salts, the decrease in current with temperature has been found to be typical of

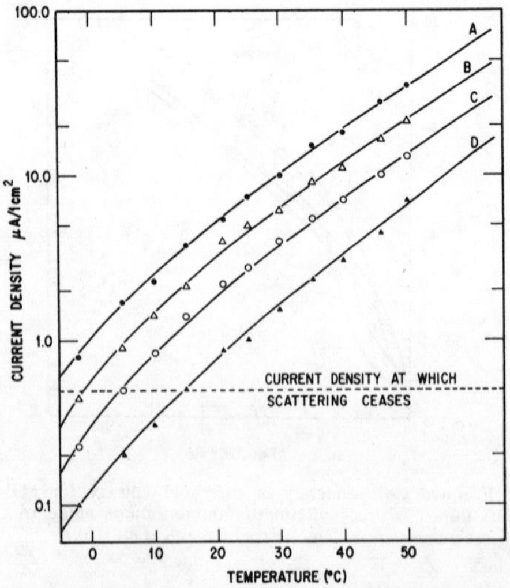

Fig. 10. AC current density as a function of temperature for MBBA doped with tetrabutylammonium perchlorate. Current density measured in 12-μm sandwich cells at 15 V_{rms}, 32 Hz, 27°C: A, 8μA/cm²; B, 5μA/cm²; C, 3μA/cm²; D, 1.4μA/cm². Dotted line indicates current density at which scattering ceased on decreasing temperatures at 15 V_{rms}, 32 Hz. (From Ristagno [43].)

electrolyte systems [43]. As a result, changing the conductive additive or the type of liquid crystal will not directly obviate the difficulty. Based on the measurements shown in Fig. 10, it is necessary to have current levels of 3 μA/cm² (27°C) to maintain display operation below 10°C. Low-temperature operation is not precluded by the nematic fluid itself, in this case, but by device power requirements. Increases in ac current with increasing temperature do not interfere with actual display operation and are only an intermittent increase in drain on the battery because watch manufacturers estimate the average operating temperature [44] to be 28°C. Producers of displays are attempting to achieve adequate battery life by several approaches: reduction of power losses in the drive system, intermittent or pulsed operation of the display, and use of more effective additives, one that will give good contrast with very low current levels.

Dynamic Scattering—DC

The earliest electrooptical work was done in displays using dc drive [4] where the principal interest was in the phenomena and not in the liquid crystal material itself. As the feasibility of practical displays came to be recognized, concern arose about the effect of passing dc current through liquid crystals, how this would affect the reliability of the display, and what the electrochemical processes were.

Materials in dc displays are continually subjected to electrochemical stress at potentials that appear sufficiently high for most organic electrochemical reactions to occur, but only recently [45] has there been much discussion of electrochemical processes in liquid crystals. Bard [46] has found the electrochemical reactions of MBBA, in solution with supporting electrolyte, to be essentially the same as those of other Schiff bases. The extension of this work into thin-cell geometry where the Schiff base itself is the solvent is not straightforward. For MBBA, studies are complicated by the need for soluble inert electrolyte and by the fact that the common impurities found in MBBA, *p-n*-butylaniline and *p*-anisaldehyde, undergo oxidation–reduction reactions at lower potentials than MBBA itself.

At present it appears that if an electrolyte such as tetraethylammonium perchlorate is added to MBBA, then the charge carrier is not MBBA itself. The species undergoing redox is not available in infinite supply because the steady-state current gradually diminishes with time and finally becomes so low that dynamic scattering no longer takes place. The amount of time for this to occur is insufficient to require significant redox of the perchlorate. During this electrolysis, changes in material composition are occurring. The nematic-to-isotropic transition temperature gradually decreases, indicating the formation of some soluble foreign species, and deposits form on the electrodes. Bubble formation has been reported [35] during the electrolysis of azoxy compounds but does not occur in MBBA. Complex electrode phenomena occur, as indicated by changes in the degree and nature of light scattering with time. All of these events affect practical display performance. As a result of changes in light scattering, current levels, and cosmetic appearance, displays driven with dc have relatively short lifetimes. A few thousand hours appears to be a reasonable limit for life at the present time, and operation above ambient temperatures accelerates failure severalfold.

Not only is the liquid crystal under electrochemical stress in dc-driven displays, but all additives and surface treatments are subjected to redox conditions. Surface-active agents that will readily undergo electrochemical oxidation or reduction should not be used in dc-driven displays. This applies to bulk surfactants and to direct surface treatments. It is evident that to have dc dynamic scattering there must be continuing charge injection, a dynamic process. For this to occur at relatively low fields requires polarization of the electrodes. The resulting double layer may be slow to form and to disperse. The accumulation of charged species at and near the electrode may interfere with the action of aligning agents, resulting in cosmetic problems.

A major problem remains in understanding dc dynamic-scattering devices: electrochemical reactions occur, but the exact sequence of reactions and their consequences need to be fully explained. A fundamental understanding of the electrochemistry of liquid crystals should lead to a better appreciation of the prospects of dc-driven devices or practical displays.

Field Effects

The basic features of materials suitable for use in field-effect displays can be ascertained from the fundamentals of the phenomena itself: an electric field is used to reorientate the molecules, and the realignment is perceived as a change in the passage of light through polarizers. Since the electrooptic effect does not rely on space-charge or current carriers, the material should be of reasonably high resistivity to minimize current flow (dc operation) or capacitive charging losses (ac). The temperature considerations are the same as discussed earlier for storage, but the effect of temperature on current can be neglected.

The equations for the threshold voltage ((3) and (4)) and response times ((1) and (2)) contain terms for Franck elastic constants, dielectric anisotropy, and viscosity.

$$V_{th} = \pi \sqrt{\frac{(K_{11} + K_{33} - 2K_{22})\pi}{\epsilon_0 \epsilon_a}} \quad \text{(see [7])} \qquad (3)$$

TABLE IV

SOME PHYSICAL CONSTANTS

Liquid Crystal	Property						Reference
	ϵ_a	$\sigma_{\|\|}/\sigma_{\perp}$	K_{11} (Newtons)	K_{22} (Newtons)	K_{33} (Newtons)	K_{33}/K_{11}	
Chlorostilbenes	−0.41	1.3					[49]
4′-cyanobenzylidene-4-methoxyaniline						1.8	[50]
Bisheptyloxyazobenzene	+0.35		[a]			1.6–2.0	[50]
PAA	−0.21[b]	1.5	9.0×10^{-12}[c]	6.0×10^{-12}[c]	21×10^{-12}[c]		[48]
MBBA (25°C)	−0.53[d]	1.5	6.10×10^{-12}		7.25×10^{-12}		[51] [52]
PEBAB	+20[e]						[53]
4′-ethoxybenzylidene-4-cyanoaniline	+17[f]						[53]
4′-hexyloxybenzylidene-4-cyanoaniline	+15[f]						[53]
4′-heptyloxybenzylidene-4-cyanoaniline	+11[f]						[53]
Licristal IV, V, VA	−0.2						[33]

[a] See [48] for elastic constants as a function of temperature for higher homologues.
[b] 120°C.
[c] 106°C.
[d] See [54] for refractive indices as a function of temperature.
[e] 125°C.
[f] Approximately 75°C.

$$V_{\text{th}} = \pi d q_0 \sqrt{\frac{\pi K_{22}}{\epsilon_0 \epsilon_a}} \qquad \text{(see [28])} \quad (4)$$

where K's are elastic constants, ϵ_a is the dielectric anisotropy, q_0 is the pitch, and d is the thickness. Equation (3) applies to nematic materials, and (4)—which includes the pitch—is applicable to the cholesteric–nematic phase transition [47]. These parameters open a whole new world to the organic chemist for material tailoring to effect fundamental changes in the display characteristics.

Perhaps the least amenable parameters are the Franck elastic constants. Relatively few of these are available in the literature, but Gruler's work on homologous azoxybenzenes [48] provides preliminary data for trend prediction. Table IV lists some known values. Unfortunately data are lacking for the (4′-ethoxybenzylidene-4-cyanoaniline) (PEBAB) family, esters, or tolans.

Dielectric anisotropy is one material characteristic that can readily be changed. Again few numbers have been published (see Table IV), but general information is available for predicting the sign of the anisotropy and trends in magnitude if not its size. The anisotropy can be made more negative by proper choice of off-axis substituents; groups with strong polarizabilities such as hydroxyl can be substituted for hydrogen and induce a change in the dielectric anisotropy of the molecule. The *ortho*hydroxyl derivative of MBBA would be expected to have a larger negative value of ϵ_a than MBBA and has been reported to be more chemically stable [55]. These benefits occur at the expense of the nematic range: 2′-hydroxyl-4′-methoxybenzylidene-4-butylaniline is nematic from 44–64.5°C. Murase [56] reports *ortho*methyl-substituted benzylideneanilines to be more chemically stable than MBBA-type Schiff bases, but gives no data on dielectric anisotropies or other material parameters.

Ortho or *meta* cyano substituents should also increase the size of the dielectric anisotropy and Gray reported work on the synthesis of these types of liquid crystals [57], but information on such materials is limited primarily to transition temperatures. Always the addition of any substituent alters the delicate balance between terminal and lateral attractions necessary for mesomorphism [31]. The general trend for PEBAB-related liquid crystals is the reduction of the size of ϵ_a by addition of terminal methylene groups [53].

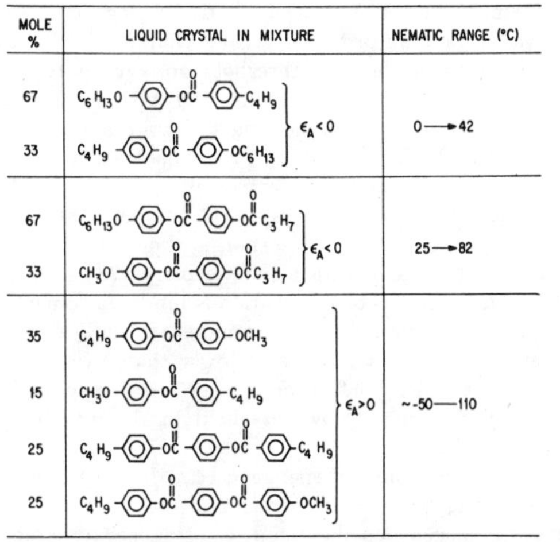

Fig. 11. Ester mixtures developed at E. Merck Company that are suitable for device applications. (From Steinstrasser [12].)

Jakeman and Raynes [28] have evaluated the effect of material parameters on threshold voltage for an electric-field-induced cholesteric–nematic transition assuming (4). It has been shown [47] that decreasing q_0 increases the decay time observed for the effect, whereas increasing ϵ_a reduces the rise time and the threshold voltage [28]. Response times were found to be minimized in materials with low viscosity, small pitch, and large dielectric anisotropy.

It should be remembered that there is no inherent reason for the accessible temperature ranges to be different in liquid crystals with positive or negative dielectric anisotropy. The concept of mixing applies to liquid crystals suitable for field-effect devices, just as it does for dynamic scattering materials. Steinstrasser [12] reported the use of esters to produce suitable materials with either positive or negative dielectric anisotropy. Some of these materials are tabulated in Fig. 11. Esters hold especial promise for device application because of increased chemical stability and white color.

For twisted nematic devices, the liquid crystal needs to have positive dielectric anisotropy. Mixing liquid crystals with positive and negative dielectric anisotropy can yield

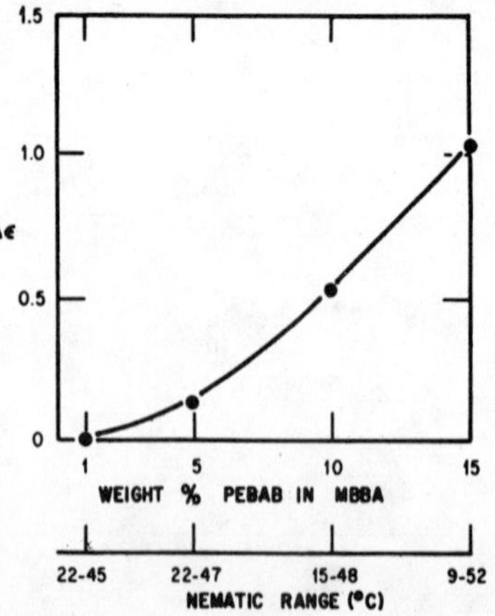

Fig. 12. Dielectric anisotropies and nematic temperatures range for PEBAB:MBBA mixtures.

room-temperature materials of positive dielectric anisotropy with values ranging from just greater than 0 to at least 5. As a result, one can choose a threshold voltage appropriate to drive conditions. For 15-percent PEBAB in MBBA, the threshold is quite sharp, occurring at 2 V_{rms}, and the display is fully on at 2.8 V_{rms}. Varying the concentration of PEBAB in MBBA can change the dielectric anisotropy (Fig. 12) and, with little change in elastic constants, can vary the threshold voltage according to (3). Response times depend on the strength of the applied field and viscosity [28], and, consequently, low-temperature operation is hindered somewhat by long response times. As in dynamic scattering, a quiescent display element responds more slowly than a recently activated segment. Response times can be reduced by maintaining the display just below threshold in the inactive or off state.

Voltage-controlled birefringence can be operated in two modes: using positive dielectric anisotropy material and initially homogeneous alignment, or using negative dielectric anisotropy and initially homeotropic alignment [8]. The latter mode is the most common and has been investigated by several groups [9] using MBBA, MBBA–Schiff base mixtures, and Merck-produced azoxybenzene materials [58].

The Future for Liquid-Crystal Displays

The prospects for devices utilizing liquid-crystal displays ultimately depend on customer acceptance, and much remains to be accomplished before the optimum device is produced. Scattering devices operated with ac drive are reliable with life limited by packaging, but dynamic scattering using dc drive fails most customer requirements. Ideally, dc displays would be filled with an electrolyte–liquid-crystal system such that all electrochemical reactions taking place would be completely reversible. The probability of approximating such a perfect system is increased by changing to field effects. In principle, dc field-effect devices can be built so that negligible current is passed, thus drastically reducing the rate of electrochemical degradation. It has not yet been established that the stringent alignment requirements can be maintained for long time periods in field-effect displays under dc or ac drive.

In all probability there will be no major improvement in operating or storage temperature range of nematic liquid crystals. The top temperature required by users of displays is attainable today, and the very low temperatures required by automobile manufacturers and the military can be satisfied by external heating. Thermal degradation of the liquid crystal can be reduced or prevented by proper packaging. Thermal stress on aligning forces or surfactants is a significant problem that has not been solved.

The response times of dynamic scattering and field-effect displays are adequate at or above room temperature for most applications. Liquid-crystal displays will not be used for certain applications until the problem of multiplexing is solved. This is being attacked by engineering of both the drive schemes and the material. The response times of cholesteric-to-nematic phase changes can be fast, but the chemist has yet to develop a material with a low threshold voltage. Significantly faster displays will come from new drive techniques rather than the liquid crystal itself unless new inherently faster electrooptical phenomena are found.

In summary, liquid-crystal displays meet the specifications of some of today's customers with regard to response times, temperature range, and reliability. Much remains to be done to establish reliability and performance under the extreme conditions of military and automotive applications. Major improvements are needed in packaging, drive circuitry, appearance, and price reduction before liquid-crystal displays will achieve their full promise.

References

[1] Y. Bjornstahl, "Investigation of anisotropic liquids," *Ann. Phys.*, vol. 56, pp. 161–207, 1918.

[2] I. G. Chistyakov, "Liquid crystals," *Sov. Phys.—Usp.*, vol. 9, pp. 551–573, 1967.

[3] J. L. Fergason *et al.*, "Display system utilizing a liquid crystal material of the cholesteric phase," U. S. Patent 3 410 999, Nov. 12, 1968.

[4] G. H. Heilmeier, L. A. Zanoni, and L. A. Barton, "Dynamic scattering: A new electrooptic effect in certain classes of nematic liquid crystals," *Proc. IEEE*, vol. 56, pp. 1162–1171, July 1968.

[5] G. H. Brown, "Liquid crystals and their roles in inanimate and animate systems," *Amer. Sci.*, vol. 60, pp. 64–73, 1972.
M. A. Berwick *et al.*, "Nematic liquid crystals for Air Force displays," Tech. Rep. AFML-TR-71-72, July 1971.
G. H. Brown, J. W. Doane, and V. D. Naft, "Structure and physical properties of liquid crystals," *Crit. Rev. Solid State Sci.*, vol. 1, pp. 303–379, 1970.

[6] L. T. Creagh, A. R. Kmetz, and R. A. Reynolds, "Performance characteristics of nematic liquid crystal display devices," *IEEE Trans. Electron Devices*, pp. 672–679, Sept. 1971.

[7] M. Schadt and W. Helfrich, "Voltage dependent optical activity of a twisted nematic liquid crystal," *Appl. Phys. Lett.*, vol. 18, pp. 127–128, 1971.

[8] F. J. Kahn, "Electric-field-induced orientational deformation of nematic liquid crystals: Tunable birefringence," *Appl. Phys. Lett.*, vol. 20, pp. 199–201, 1972.

[9] M. F. Schiekel and F. Fahrenschon, "Deformation of nematic liquid crystals with vertical orientation in electrical fields," *Appl. Phys. Lett.*, vol. 19, pp. 391–393, 1971.

[10] U. Bonne and J. P. Cummings, "Nematic liquid crystal displays: Properties and limitations," presented at the 1972 IEEE Conf. Display Devices (New York, Oct. 1972).

[11] V. Freederickz and V. S. Zolina, "Forces causing the orientation of an anisotropic liquid," *Trans. Faraday Soc.*. vol. 29, pp. 917–930, 1933.

[12] R. Steinstrasser, "Nematic liquid crystalline *p*, *p'*-disubstituted phenyl benzoates," presented at the 4th Int. Liquid Crystal Conf. (Kent, Ohio, Aug. 1972), Paper 119.

[13] H. Kelker, B. Scheurle, R. Hatz, and W. Bartsch, "Liquid crystalline phases having particularly low freezing points," *Angew. Chem.* (Int. ed.), vol. 9, pp. 962–963, 1970.

[14] D. J. Giancola, General Manager, Princeton Organics, private communication.

[15] L. T. Creagh and A. R. Kmetz, "Selective doping of MBBA to optimize display performance," *J. Electron. Mater.*, vol. 1, pp. 350–354, 1972.

[16] H. Kelker and B. Scheurle, "A liquid crystalline (nematic) phase with a particularly low solidification point," *Angew. Chem.* (Int. ed.), vol. 8, pp. 884–885, 1969.

[17] E. F. Carr, "Anomalous alignment in the smectic phase of a liquid crystal owing to an electric field," *Phys. Rev. Lett.*, vol. 24, pp. 807–

809, 1970.
[18] J. L. Haberfeld, E. C. Hsu, and J. F. Johnson, "Liquid crystal purification by zone refining," presented at the 4th Int. Liquid Crystal Conf. (Kent, Ohio, Aug. 1972), Paper 171.
[19] N. J. Felici, "The de-ionization of strongly polar liquids," *Brit. J. Appl. Phys.*, vol. 15, pp. 801–805, 1964.
[20] G. Briere, N. Felici, and J. C. Filippini, "Deionisation du nitrobenzene par electrodialyse," *C. R. Acad. Sci.* (Paris), vol. 261, pp. 5097–5099, 1965.
[21] G. Briere and F. Gaspard, "Relaxation of photoconduction in nitrobenzene," *Chem. Phys. Lett.*, vol. 7, pp. 537–540, 1970.
[22] R. Tobazeon, J. C. Filippini, J. Borel, J. Robert, and X. Poggi, "Nematic liquid crystal purification by electrodialysis. Behavior of MBBA in electric fields," presented at the Int. Liquid Crystal Conf. (Berlin, Aug. 1970).
[23] R. A. Kashnow and H. S. Cole, "Electrohydrodynamic instabilities in a high-purity nematic liquid crystal," *J. Appl. Phys.*, vol. 42, pp. 2134–2135, 1971.
[24] L. T. Creagh, "Electro-optical display devices using nematic mixtures with very wide temperature ranges," U. S. Patent 3 655 270, Apr. 11, 1972.
[25] S. E. B. Petrie, "Low temperature thermal behavior of liquid crystalline materials," presented at the Phys. Dem. Div. ACS 162nd Nat. Meeting (Washington, D. C., Sept. 1971), Paper 210.
[26] ——, private communication.
[27] M. J. S. Dewar and J. P. Schroeder, "Liquid crystals as solvents I. The use of nematic and smectic phases in gas–liquid chromatography," *J. Amer. Chem. Soc.*, vol. 86, pp. 235–239, 1964.
[28] E. Jakeman and E. P. Raynes, "Electro-optic response times in liquid crystals," *Phys. Lett.*, vol. 39A, pp. 69–70, 1972.
[29] H. Koelmans and A. M. van Boxtel, "Electrohydrodynamic flow in nematic liquid crystals," *Phys. Lett.*, vol. 32A, pp. 32–33, 1970.
[30] D. Berchet, A. Hochapel, and R. Viovy, "Variation de la viscosite de la phase mesomorph nematique en function de la temperature," *C. R. Acad. Sci.* (Paris), vol. 270, pp. 1065–1068, 1970.
[31] G. W. Gray, *Molecular Structure and the Properties of Liquid Crystals.* New York: Academic Press, 1962, pp. 139–144.
[32] W. Maier and A. Saupe, "Eine einfache molekular-statistische Theorie der nematischen Kristallin flüssigen Phase. Teil II," *Z. Naturforsch.*, vol. 15A, pp. 287–292, 1960.
[33] E. Merck, data sheet on Nematic Phase 4, 5, 5A, and 6, Darmstadt, Germany, 1972.
[34] *Handbook of Chemistry and Physics.* Cleveland, Ohio: Chemical Rubber, 1958.
[35] L. Pohl, R. Steinstrasser, and B. Hampel, "Performance of nematic phase 5, 5A, and 6 in liquid crystal displays," presented at the 4th Int. Liquid Crystal Conf. (Kent, Ohio, Aug. 1972), Paper 144.
[36] I. Haller, "Ultraviolet protecting coating," *IBM Tech. Disclosure Bull.*, vol. 15, pp. 584–585, 1972.
[37] W. Helfrich, "Conduction-induced alignment of nematic liquid crystals: Basic model and stability considerations," *J. Chem. Phys.*, vol. 51, pp. 4092–4105, 1969.
E. F. Carr, "Ordering in liquid crystals owing to electric and magnetic fields," in *Ordered Fluids and Liquid Crystals.* Washington, D. C.: Amer. Chem. Soc., 1967, pp. 76–88.
[38] N. Felici, "Phenomenes hydro et aerodynamiques dans la conduction des dielectriques fluides," *Rev. Gen. Elec.*, vol. 78, pp. 717–734, 1969.
[39] W. H. DeJeu, C. J. Gerritsma, and A. M. van Boxtel, "Electrohydrodynamic instabilities in nematic liquid crystals," *Phys. Lett.*, vol. 34A, pp. 203–204, 1971.
[40] Orsay Liquid Crystal Group, "AC and dc regimes of the electrohydrodynamic instabilities in nematic liquid crystals," *Mol. Crys. Liq. Crys.*, vol. 12, pp. 251–266, 1971.
[41] S. Stein and R. A. Kashnow, "A two-frequency coincidence addressing scheme for nematic liquid crystal displays," *Appl. Phys. Lett.*, vol. 19, pp. 343–345, 1971.
[42] C. V. Ristagno, private communication.
[43] C. T. Baker and I. Trachtenberg, "Conductivity of KOH and KOH–ZnO electrolytes from 36° to −66°C," *J. Electrochem. Soc.*, vol. 144, pp. 1045–1046, 1967.
[44] J. McDermott, "How good are the new watches? Electronic claims split industry," *Electron. Des.*, vol. 20, pp. 30–32, 1972.
[45] A. Sussman, "Electrochemical degradation in mesomorphic devices," presented at the 4th Int. Liquid Crystal Conf. (Kent, Ohio, Aug. 1972), Paper 149.
[46] A. Lomax, R. Hirasawa, and A. Bard, "The electrochemistry of the liquid crystal n-(p-Methoxybenzylidene)-p-n-butylaniline (MBBA)," *J. Electrochem. Soc.*, vol. 119, pp. 1679–1682, 1972.
[47] J. J. Wysocki, J. Adams, and W. Haas, "Electric-field-induced phase change in cholesteric liquid crystals," *Phys. Rev. Lett.*, vol. 20, pp. 1024–1025, 1968.
[48] H. Gruler, "Investigations of the elastic constants of the nematic homologous series of 4,4′-(n-Alkoxyazobenzene)," preprint.
[49] J. Sprockel, private communication.
[50] H. Gruler and G. Meier, "Electric field-induced deformations in oriented liquid crystals," *Mol. Crys. Liq. Crys.*, vol. 16, pp. 299–310, 1972.
[51] D. Dignet, F. Rondelez, and G. Durand, "Anisotropie de la constante dielectrique et de la conductivite du p-methoxy-benzylidene-p-n-butyl-aniline en phase nematique," *C. R. Acad. Sci.* (Paris), vol. 271 G, pp. 954–957, 1970.
[52] P. A. Penz, "Electromagnetic hydrodynamics of liquid crystals," *Phys. Rev. A*, vol. 6, pp. 414–425, 1972.
[53] M. Schadt, "Dielectric properties of some nematic liquid crystals with strong positive dielectric anisotropy," *J. Chem. Phys.*, vol. 56, pp. 1494–1497, 1972.
[54] M. Burnet-Germain, "Indices du p-methoxybenzylidene p-n-butylaniline (MBBA)," *C. R. Acad. Sci.* (Paris), vol. 271 B, pp. 1075–1077, 1970.
[55] I. Teucher, C. M. Paleos, and M. M. Labes, "Properties of structurally stabilized anil-type nematic liquid crystals," *Mol. Crys. Liq. Crys.*, vol. 11, pp. 187–189, 1970.
[56] K. Murase, "Liquid crystalline properties of laterally substituted benzylideneaniline derivatives," *Chem. Lett.* (Chem. Soc. Japan), pp. 471–472, 1972.
[57] W. W. Gray, "Recent advances in preparations of liquid crystals," presented at the 4th Int. Liquid Crystal Conf. (Kent, Ohio, Aug. 1972), Plenary lecture P-8.
[58] W. Greubel and U. Wolff, "Electrically controllable domains in nematic liquid crystals," *Appl. Phys. Lett.*, vol. 19, pp. 213–215, 1971.
[59] J. A. Castellano, J. E. Goldmacher, L. A. Barton, and J. S. Kane, "Liquid crystals II. Effect of terminal group substitution on the mesomorphic behavior of some benzylideneanilines," *J. Org. Chem.*, vol. 33, pp. 3501–3504, 1968.
[60] Houben–Weyl–Muller, *Methoden der Organischen Chemie*, vol. 10. Stuttgart, Germany: Thiem, 1965, p. 762.
[61] S. A. Haut, D. C. Schroeder, and J. P. Schroeder, "Liquid crystals II. Unsymmetrical p-phenylene di-p-alkoxybenzoates," *J. Org. Chem.*, vol. 37, pp. 1425–1428, 1972.
[62] W. R. Young, A. Aviram, and R. J. Cox, "New, nonplanar *trans*-stilbenes exhibiting nematic phases at room temperature," *Angew. Chem.* (Int. Ed.), vol. 10, pp. 410–411, 1971.
[63] J. Malthete, M. Leclercq, J. Gabard, J. Billard, and J. Jacques, "Tolans nematiques," *C. R. Acad. Sci.* (Paris), vol. 273, pp. 265–267, 1971.
[64] J. van der Veen, W. H. De Jeu, A. H. Grobben, and J. Boven, "Low melting liquid crystalline p, p′-di-n-alkyl-, axoxy-, and azobenzenes," *Mol. Crys. Liq. Crys.*, to be published.
[65] J. P. Schroeder, D. C. Schroeder, and M. Katsikas, "Stationary liquid phases in gas–liquid chromatography," in *Liquid Crystals and Ordered Fluids.* New York: Plenum, 1970, p. 178.
[66] S. L. Arora, J. L. Fergason, and T. R. Taylor, "Molecular structure and liquid crystallinity. Phenylene bis(alkoxybenzoates)," *J. Org. Chem.*, vol. 35, pp. 4055–4058, 1970.
[67] M. J. S. Dewar and R. S. Goldberg, "Effects of central and terminal groups on nematic mesophase stability," *J. Org. Chem.*, vol. 35, pp. 2711–2715, 1970.
[68] S. L. Arora, J. L. Fergason, and A. Saupe, "Two liquid crystalline phases with nematic morphology in laterally substituted phenylenediame derivatives," *Mol. Crys. Liq. Crys.*, vol. 10, pp. 243–257, 1970.

Self-checking Quiz

Each of the questions is followed by a reference to a section or table of the text where the material is explained. For maximum benefit from these questions the reader should attempt a solution before rereading the referenced material.

1. What is the difference between light-emitting and light-controlling displays? What is the main factor that is different between light-emitting and light-controlling displays? *Ans.:* See Gordon and Anderson, Section II.
2. Display devices can be classified in the following chart:

	Beam-Addressed	Matrix-Addressed
Light-emitting		
Light-controlling		

   Locate the following types of displays in this chart.
   cathode-ray tube (CRT)
   light emitting diode (LED)
   liquid crystal
   plasma
   eidophor
   Ans.: See Gordon and Anderson, Sections III and IV.
3. Compare the advantages and disadvantages of AC and DC plasma displays, LED displays, and liquid crystal displays in the indicator and small display area. *Ans.:* See Gordon and Anderson, Sections III and IV and Tables IV, V, VII, and IX.
4. With integrated circuits, new technologies (CMOS, I^2L) are introduced commercially relatively rapidly after the first research devices are developed. With

light-emitting diodes, gallium arsenide phosphide, and gallium phosphide, devices remain dominant commercially although others have performance characteristics that are desirable. Why? *Ans.:* See Gordon and Anderson, Section V, and Craford and Groves, Section II.

5. Compare the electroluminescent properties of various types of VPE $GaAs_{1-x}P_x$ and GaP LEDs in terms of quantum efficiency and luminous power efficiency. Which parameter is most important in display applications? *Ans.:* See Craford and Groves, Section V and Table II.
6. Compare LPE and VPE LED technology and capabilities. *Ans.:* See Craford and Groves, Sections VII and VIII.
7. What is the operating temperature range of commercially available nematic liquid crystals? How do these compare to the temperature range required in most applications? *Ans.:* See Creagh, Tables I and II.
8. Most commercially available liquid crystal displays have used "dynamic scattering," although increasing use is being made of various field effects. Why? *Ans.:* See Creagh (last half of article) and Gordon and Anderson, Section IV:

Index